Dedicated to **Agriculture**

Humankind's Greatest Innovation

Review: *Ecosystem-based Agriculture: The Pillar of Global Food Security* by Professor Trisit Chakraborty and Professor George Bird advances the concept of ecosystem-based agriculture, which is an ecologically compatible yet practically feasible method of agriculture that allows food production without compromising natural soil fertility and endangering soil biodiversity.

The authors have adopted a realistic approach towards food production while keeping the question of ecological sustainability in the fore. It is to be realized that it may not be possible and practicable to practice ecosystem-based agriculture in its classical form despite its contribution to ecosystem sustainability, rural employment generation and a higher quality of life. In this regard, Chakraborty and Bird have taken a pragmatic stand where they recommend *Situation Specific Technological Support* for focusing on biological-based pest management, organic methods, crop rotation, biological diversity maintenance and other ecological goals. This flexible yet focused approach contributes significantly to the applicability and acceptability of the agricultural strategies advocated by the authors.

The authors also suggest a phase-wise shift to ecosystem-based agriculture by first implementing a modified two-system approach where some of the land is dedicated to soil health renovation and biological diversity enhancement, while the other part is allowed to employ conventional methods. This could for the time being achieve a balance between conservation and production.

The book is a pleasure to read because of the simple yet comprehensive manner in which it has chronicled the history of agriculture since its inception about 10,000 years ago, and then started flourishing in the valleys of Nile, Tigris and Euphrates and Indus, to eventually spread all over the world. It also shows that these agricultural systems supported local crop diversity, which was later reduced drastically by the homogenizing effects of conventional agriculture. Hence, the prescription of ecosystem-based agriculture advanced by the authors – albeit in a new avatar – is a significant contribution of use to academics and agricultural practitioners.

Abhik Gupta, Vice Chancellor,
Assam University
Silchar, India

Table of Contents

Prologue

Since the formation of the Universe about 14 billion years ago and Earth about 9.5 billion years later, our planet has undergone at least ten major changes (Fig. 1). In each of these phases there have been numerous epics, eras, famines and pandemics.

Figure 1. 4.5 billion-year history of the planet Earth.

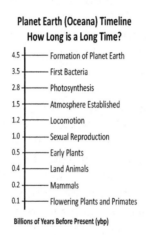

Planet Earth (Oceana) Timeline
How Long is a Long Time?

4.5	Formation of Planet Earth
3.5	First Bacteria
2.8	Photosynthesis
1.5	Atmosphere Established
1.2	Locomotion
1.0	Sexual Reproduction
0.5	Early Plants
0.4	Land Animals
0.2	Mammals
0.1	Flowering Plants and Primates

Billions of Years Before Present (ybp)

One thing that is absolutely certain is that change will continue. The future will be different than the past. In a world of more than 7.8 billion people, with 4.5 billion living in Asia with limited natural resources (clean air, fresh water, healthy soil and biological diversity), it is clear that human ingenuity will be essential for providing the agriculture essential for humankind to maintain an appropriate quality of life. Today, about 800 million people are undernourished as a result of inadequate food availability. Millions more suffer various maladies as a result of poor-quality food. A new era of sustainable and equitable agriculture based on the Fundamentals of Ecology is imperative for a high quality of life for future generations.

Professor Chakraborty: Like many Indians, I belong to an average Bengalee family living in a village, named Joychandi, now part of Bangladesh. In pre-independence days, we lived together under a joint family system in a single cooking arrangement as per our ancestral culture. A joint agricultural farm with a rice-based cropping system and kitchen gardening was the main source of income of the family. My father and two uncles also worked in tea gardens. Another uncle practiced homeopathic medicine in a local town, in addition, to jointly supervising our farm with my grandfathers.

I was born in 1939. As the youngest son of my parents, I was brought up traditionally in the midst of a natural habitat during the difficult food crisis days after the end of World War II. I enjoyed the simple village life and tea garden atmosphere, along with learning about food production. During my childhood, while staying at my parent's house, I observed various aspects of agriculture, including how to plough, uproot rice seedlings, plant rice seedlings in muddy fields and use of bamboo traps to catch small fish in stagnant rice fields. I enjoyed playing and running around piles of rice sacks kept for threshing by a pair of bullocks. A bamboo ladder was pulled behind as the traditional way of rice threshing. I also learned to catch fish using a simple and narrow, but strong, bamboo stick. This traditional way of fishing was used in local ponds and other bodies of water. Taking care of our ducks and feeding green grass to the cows were part of my daily chores.

After the partition of India, I left my village home and went to Munshibajar. I lived at a tea garden and was admitted to a Middle English school. Because of political disturbances, I had to leave school and move to India, leaving my homeland behind. My parents admitted me to Doom Dooma Hoonlal High School, Assam where I stayed in a tea garden with my eldest sister until tenth grade. There I observed

7

every component of tea cultivation, from plucking leaves by female laborers to final processing. This background influenced me to choose agriculture for my education and life-long career in teaching and research. I received my Master of Science and Doctor of Philosophy degrees in agronomy, specializing in weed science, grassland management and low cost phospho-organic farming in rice-based cropping systems.

It is interesting to note that both Professor Bird and I came from similar natural and agricultural backgrounds. We both spent our early childhood days in the midst of traditional systems of agriculture. This becomes our added advantage as we write this book for the welfare of the people of both the East and West.

Professor Bird: By shortly after my sixth birthday, my family had moved four times and I had lived in five different communities in the U.S.A. Because of this disruption at a young age, I began to feel most comfortable at my Great-Uncle's farm in Simonsville, Vermont; a place where I spent my summers. It was a large and modern poultry farm for the time, but also had dairy cows, Nubian goats, numerous hobbies and served as a Guest Farm for vacationers from New York City. Simonsville is halfway between Chester and Londonderry and small enough to have had its Post Office closed in 1949. I never missed morning milking. I soon learned to take the cows to pasture after milking and bring them back to the barn in the afternoon. I also learned to shoot fox on the chicken range. During haying season, I was allowed to rake hay with a one-horse dump rake. Prince, a small work horse, was my favorite. I rode Prince on old Vermont logging trails. We even practiced a few small jumps. It was in the forests of Vermont that I fished in brooks with my simple pole, most likely very similar to those used by my ancestors when they moved west in their covered wagon. I would often encounter bear feeding on blackberries. I only got lost once. It was a long walk back to the farm, but really nothing compared to those of my pioneer ancestors.

Like Trisit, I was born in 1939. My father Joe was born in Indian Territory and his mother in a log cabin on the Kansas prairie. Because of these and other farm experiences, I chose agriculture for my career. My B.S., M.S. and Ph.D. were in horticulture, nematology/entomology and phytopathology, respectively.

The objective of *Ecosystem-Based Agriculture: The Pillar of Global Food Security* is three-fold: 1) to identify and describe the pillars of ecosystem-based agriculture as they relate to global food security and zero poverty/zero-hunger, 2) to provide the biological/ecological background essential for understanding how the pillars function and why they are essential and 3) to provide several policy recommendations for the government of India.

The book includes descriptions of how food systems evolved, fundamentals of nature and science, domain of eco-literacy, and concept of nature-loving/ecosystem-based food systems. These topics are supported with descriptions of the concepts of sustainable and equitable development, agricultural innovations, pest and disease prevention and management, global hunger/food security, rice production, and challenges related to the future of humankind.

Ecosystem-based agriculture is imperative at a time when reductionist science naively believes that humans can have complete control of Earth through implementation of the concepts of an Anthropocene (Crutzen, 2000). Based on the fundamental Laws of Science and Nature, there is no evidence that an Anthropocene can be successful. On the contrary, the future of an appropriate quality of life for humankind mandates optimal interactions between an ecosystem-based agriculture and the clean air, freshwater security, healthy soil, biological diversity and human ingenuity upon which it depends. Fortunately, in 2019, Boehnert proposed an additional alternate future epoch: the Ecocene.

9

Much has been written about the general nature of agriculture. The scientific literature is also rich with fundamentals about how living systems work. Much more will be written in the near future as the topics of local and global food systems expand. It has been Professor Bird's experience in academia, however, that eco-literacy is a common deficiency. This may explain why there are relatively few works describing the ecosystem basis of agriculture in terms that do not require a comprehensive understanding of the language of science. Professor Chakraborty's vast experience with agriculture in India provides a highly unique opportunity to demonstrate why the sub-continent of India is the ideal place to study the history of agriculture. The sub-continent of India represents one of, if not, the oldest known continuous societal histories of humankind. Its geographical regions range from relatively close to the equator to more than one third of the way to the north pole. Elevations range from sea level to more than 8,500 meters. When this diversity is combined with the fact that India currently has more than 17% of the earth's people, it makes India an excellent, if not the best, place to access the past, present and future of agriculture! The authors strongly believe that *Ecosystem-Based Agriculture: The Pillar of Global Food Security* provides an appropriate road map for small and large farms in high-income and low-income nations.

A famous philosopher once said, *if you wish to communicate with me, kindly define your terms.* In concert with this recommendation, brief glossaries of pertinent terms are included in some of the chapters.

Chapter 1. Pillar of Global Food Security

- *A system is something with two or more interactive parts.*
- *Ecosystem-based agriculture is a sustainable system of food, feed and fiber production based on solar energy, healthy soil and nature-loving plant species. It is significantly enhanced when animals are included in the system.*
- *Healthy soils are biologically active and respond to management in a predictable manner.*
- *Nature-loving plant species are uniquely adapted to the specific temperature, moisture and microbial regimes of their origins.*

Life on our planet evolved over the past 3.5 billion years. It was the result of interactions between matter (stuff) and energy (ability to do work). It is governed by the fundamental laws of science and nature. Agriculture is no exception. Food systems require a multitude of different living organisms to function. These organisms are divided into three Domains and at least twenty-three Kingdoms based on their genetics, appearance, habitats and behavior (Figure 1.1). They have the ability to replicate, take in matter and energy from external sources, respond to their environment and expel unused matter and energy. All organisms require energy and matter for their growth, metabolic processes and reproduction.

Early humans survived primarily as hunters and gathers. Following a global retreat of glaciers, the environment was right for development of humankind's greatest innovation: agriculture! This monumental process/event led to many significant changes in society.

Agricultural systems are complex and contain multiple interactive parts. They are open systems requiring matter (solid, liquid and gas) and energy from external sources. The outputs of agriculture have emergent properties that are not present in their component parts. Primary production through the process of photosynthesis in green plants forms the basis of food systems. Agriculture is performed with assistance from interactions among many types of living organisms. As a result of human innovations, the nature of agriculture has changed greatly over time. Because of its geographical diversity, history and large population, the sub-continent of India is well suited for studying these changes.

Figure 1.1 Three Domains and 23 Kingdoms of life, plus viruses and prions.

3 Domains and 23 Kingdoms of Life
(plus viruses and prions)

Eukarya	Arachaea	Bacteria	Prions	Viruses
• Animals • Plants • Fungi • Flagellates • Water Molds • Slime Molds • Ciliates • Entamoebae • Trichomoads • Microporidi • Diplomonads	• Six Kingdoms	• Six Kingdoms	• Protein Fragment	• ssRNA • dsRNA • ssDNA • dsDNA • Multiple Component Viruses

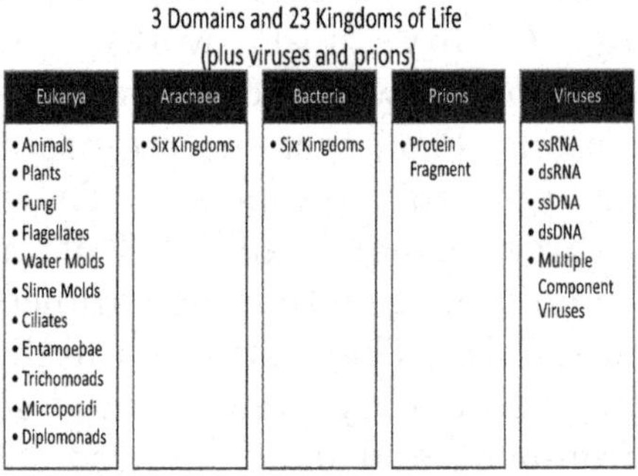

Different types of agriculture require different external material, energy and human inputs. As new systems evolve, they should be designed and managed to optimize productivity close to the system's dynamic equilibrium. If the growth of a system excessively exceeds its basic resources, the system is likely to senesce and possibly collapse. This takes place when the system reaches an unanticipated bifurcation (tipping) point. Throughout history, the decline and collapse of agricultural systems have taken place many times in many different places. This can result in famine or significant human migrations. The causes of these tragedies are complex. They may be the results, among others, of pestilence, disease, excessive soil health degradation or disruption by war.

Today, there are many types of agricultural systems that serve the needs of individuals, families, local communities, nations, corporations or the global community of humankind. Unfortunately, many of these systems are not compatible with the long-term sustainability of the resources required for their success. To rectify these potentially negative situations, it is imperative to have a fundamental understanding of how food systems work. This should include knowledge about the following eleven topics related to ecosystem-based and nature-loving food systems: 1) food system history, 2) the science of how the world is known to work, 3) eco-literacy, 4) types of farming systems, 5) integrated pest and disease management, 6) agricultural innovations, 7) fundamentals of sustainable and equitable development, 8) global hunger and food security, 9) role of rice, 10) global status of ecosystem-based food systems, and 11) future challenges and strategies related to food security and zero-hunger. At maturity, ecosystem-based and nature-loving feed, food and fiber systems are self-regulating. This occurs in a manner that reduces risk of the system exceeding the limits of its dynamic equilibria. These views from western science are highly compatible with ancient perceptions of nature.

The six pillars that support the foundation of ecosystem-based agriculture include: 1) healthy soil, 2) ecological harmony, 3) production technologies, 4) integrated pest and disease management, 5) support systems and 6) market networks/standards (Fig. 1.2). All of these are essential for success. The pillar referred to as support systems includes

14

the domains of research, technology development, and education. While it is recognized that marketing networks and standards are essential for success, they are not covered in a major way in this book.

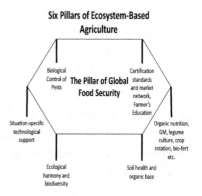

Figure 1.2. The six pillars of ecosystem-based agriculture as the overall support system for global food security.

In most cases, implementation of successful ecosystem-based agriculture requires change. There are three primary tools available for implementing change. These are education, facilitation and persuasion (Rogers and Shoemaker, 1971; Zaltman and Duncan, 1977). The authors are well aware that education can be a difficult process and should be focused on the 12.5% percent of the population known as early adopters. The early adopters will be the primary educators for the 35% of the population known as the early majority. Facilitation consists of the items necessary for adoption of a proposed change. If these are not readily available, it is highly probable that the desired change will not take place. Persuasion is the third tool available to assist in change. It may be implemented as either a "carrot" (reward) or a "stick" (punishment). In 1995, Brooks published the third edition of his land-mark treatise about challenges associated with implementation of innovations.

Chapter 2. Nature and Science

Systems are entities with two or more interactive parts. Nature is a complex system with many interactive parts. Nature is an open system. It requires energy from an external source, the sun, to transform matter into its interactive parts. The parts are dynamic and forever changing. As an open system, nature gives off residuals (never wastes!) that are incorporated into other essential parts of the system or released into the external environment. The components of nature respond to their immediate environment. In addition, living systems have the capability of replicating themselves. Based on these fundamental attributes, nature is a dynamic living system. Today, there are very few, if any, truly natural systems left on our planet. By far, the vast majority of ecosystems have been impacted in one way or another from the activities of humankind.

For the vast majority of the past 12,000 years, we (humans) have lived in the Geological Time Period (Epoch) known as the Holocene. During the last 300 years, technologies evolving from the western Industrial Revolution have had major impacts on the ecology of our planet. As a result, it has been proposed that we now live in a Human Epoch known as the Anthropocene (Crutzen, 2000). More recently, Boehnert (2019) proposed the Ecocene as an alternate future. To achieve a successful Ecocene, the impacts of our innovations must be long-term *blessings and not curses* to humankind (paraphrased from Albert Einstein). They must foster the evolution of traits necessary for a desirable quality of life for future generations of the people of our planet. For this to take place, it is imperative for science, technology, the humanities and food systems to work in harmony with each other.

- *Science.- Science is a body of knowledge and the requisite methodology used to acquire new understandings about how the world works. As a process, science involves*

18

- *development and testing of alternate hypotheses. It is important to recognize, however, that there are ways to acquire knowledge other than through scientific method.*

- *Technology.- Technology is applied knowledge. As a process, it involves the engineering and construction of useful innovations.*

- *Humanities.- Humanities consist of activities related to the human condition. They include history, language, religion, art, music, philosophy, law and many others which impact the thoughts and actions of society.*

- *Agriculture.- Agriculture is the process of producing food, feed and fiber for human or animal use.*

Western science had its beginnings in Greek times through the discoveries of Thales who was born about 640 B.C. By the time of Aristotle, born in 384 B.C., nature was understood well enough for him to have a holistic philosophy indicating that structure (matter/form) interacts with processes to result in every-changing patterns. This is very similar to the present concept of complex open systems associated with current holistic science and holistic thinking.

Greek science ceased around 640 A.D. It was reborn at the end of the Dark Ages, with modern western science beginning in the Sixteenth Century. It was during this interval that the great Bubonic Plague pandemic reached Europe. The works of Galileo and Descartes gave rise to the concept of reductionism, the belief that a system can be understood and regulated through an understanding of its parts. This forms the concept of linear systems. Although it remains as the dominant concept basis of science and technology, it is now recognized that nature and agriculture are open systems that have emerging properties not present in their parts (Wessels, 2013). These complex non-linear systems are also known to consist of nested subsystems, self- organizing attributes and have the potential to change rapidly in an unanticipated or non- predictable manner upon reaching a bifurcation (tipping) point. Agriculture is a complex non-linear system. COVID-19 is also a complex non-linear system. As an infectious disease epidemic, it reached a bifurcation point early in 2020.

Our institutions (academia, government, private business and agriculture) are not well-designed to understand and properly foster the direct and indirect multi-trophic interactions among the domains of science, technology, the humanities and agricultural systems. This will have to change for widespread adoption of ecosystem-based and nature-loving agriculture.

Chapter 3. Eco-Literacy

Fundamentals

Eco-literacy is imperative for understanding global food security. Understanding the fundamentals of ecology requires an elementary knowledge of physics, chemistry and biology. The goal should be for humankind to communicate and think in the language of ecology. The following concepts form the basis for eco-literacy.

- *Energy.- The ability to do work as measured in calories, the amount of heat required to raise one gram of water one degree Celsius.*

- *First Law of Thermodynamics.- Energy is neither created or destroyed. Energy is transformed from one type to another. This concept also applies to the Law of Conservation of Matter.*

- *Second Law of Thermodynamics.- When energy is transformed from one type to another, some of the system's energy is lost. This is known as the Law of Entropy.*

- *Atom.- The smallest electrically neutral complete particle of matter. Atoms contain subatomic particles such as protons, neutrons and electrons. Atoms have an equal number of protons (positive charged subatomic particles) and electrons (negatively charged subatomic particles).*

- *Ion.- The smallest complete electrically charged particle of matter. Ions have either fewer electrons than protons as positively charged ions or more electrons than protons as negatively charged particles. Ions with unlike charges attract each other to form ionic bonds; whereas, ions with like charges repel each other.*

- *Chemical Compound.- A chemical substance formed from the union of two or more different chemical elements. If it is carbon-based, it is an organic substance. If it is like table salt with ionic bonding, it is an inorganic substance.*

- *Plant or Animal Cell.- The smallest complete unit of plant or animal tissue.*

- *Mitosis.- The process of cell division resulting in two identical daughter cells required for one of the two processes of growth. As a second process, growth can take place when individual cells enlarge.*

- *Meiosis.- The process of cell division resulting in the formation of gametes (sperm or eggs) required for sexual reproduction.*

- *Ecosystem.- A place where biotic (living organisms) interact with their abiotic (non-living) environment.*

The science of ecology serves as the backbone of agriculture as it relates to sustainable/equitable development and global food security. Meadows, Meadows and Randers (1992) defined development as the qualitative process of bringing something to a fuller or better state. Development requires moral/ethical value judgements. It is different than growth. Growth is a quantitative process resulting in increased mass/size. Odum, an early American ecologist, defined ecology as the relationship between the environment and biotic world as it relates to the structure and function of nature. The word ecology originates from the Greek word, Oikos, meaning abode or house of organisms and popularly known as habitat. As a functional limit of habitat, Joseph Grinnell coined the term, *nische* in 1917 to denote the microhabitat where a species, population or individual lives. Thus, in ecosystem-based agriculture in lowlands where deep water rice is grown along with fish and ducks on the surface, all of these organisms have different niches. The concept of niche provides an indication of how a species interacts in harmony with the rest of its ecosystem.

Energy is distributed throughout ecosystems in accordance with the first and second laws of thermodynamics as they relate to the processes of photosynthesis (Fig. 3.1) and respiration (Fig. 3.2). Matter is either carbon-based electrically neutral organic matter or inorganic matter formed from the bonding of ions of unlike charges. These two types of matter transition from one to another as new organic matter is formed in the process of photosynthesis, or when old organic matter undergoes the process of decomposition.

Atoms and ions of many different types of elements are essential for the development of plant tissue. In general, plant roots take-up elements in their ionic forms. In soil, most of the atoms of these elements are bound to carbon through co-valent bonds and not available for direct uptake by plant roots. This matter is often referred to as fixed, sequestered or unavailable. For example, nitrogen which is an essential component of protein and nucleic acids, resides in the bodies of living bacteria. For the atoms of these elements to be available for uptake, they must be mineralized and converted into their ionic forms (Fig. 3.3).

Figure 3.1. The process of photosynthesis is energy dependent. It converts carbon dioxide and water into glucose in the presence of energy, forming the basis of terrestrial life.

Photosynthesis
(An energy-demanding process)

$$6CO_2 + 12H_2O \longrightarrow C_6H_{12}O_6 + 6O_2 + 6H_2O$$

Carbon Dioxide (gas)   Water   Light Energy Input   Glucose   Oxygen (gas)   Water

(Low-Quality Potential Energy)   (High-Quality Potential Energy)

Figure 3.2. Respiration is an energy-releasing process. In the presence of oxygen, high energy potential glucose is burned, releasing energy and converting the glucose into low potential energy to carbon dioxide and water.

Cell Respiration
(An energy-releasing process)

$$C_6H_{12}O_6 + 6O_2 \longrightarrow 6CO_2 + 6H_2O$$

Glucose   Oxygen   Energy Released   Carbon Dioxide   Water

(High-Quality Potential Energy)   (Low-Quality Potential Energy)

Figure 3.3. Processes of mineralization (conversion of part of an organic compound to an ion) and fixation (attachment via covalent bonding of part of an ion to an organic compound).

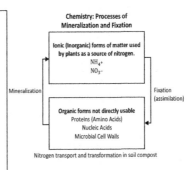

Chemistry: Processes of Mineralization and Fixation

Ionic (Inorganic) forms of matter used by plants as a source of nitrogen.
NH_4^+
NO_3^-

Mineralization

Fixation (assimilation)

Organic forms not directly usable
Proteins (Amino Acids)
Nucleic Acids
Microbial Cell Walls

Nitrogen transport and transformation in soil compost

A group of individuals of the same species is referred to as a population. Communities are comprised of groups of populations. An ecosystem is composed of populations and communities interacting with their abiotic environment (Fig. 3.4). Ecosystems may be terrestrial or aquatic (Fig. 3.5). Terrestrial ecosystems are regulated primarily by temperature and water. Large ecosystems such as a cold desert or a hot rainforest are called biomes. The transition area between an aquatic and terrestrial ecosystem is an ecotone. The largest biome is the biosphere, the thin layer of life surrounding the earth's surface.

Ecosystems are composed of trophic groups of organisms that obtain the matter and energy necessary for their life processes. The primary producers undergo photosynthesis. Herbivores feed on primary producer tissue. Carnivores feed on herbivores, omnivores or other carnivores.

Trophic groups are organized into levels represented by a pyramid. The first or basal trophic group consists of autotrophs, the green plants that undergo the process of photosynthesis. As self-feeders, they obtain their energy from the sun and atoms necessary for production of glucose (an organic compound) from atmospheric carbon dioxide and water (Fig. 3.6 and Fig. 3.7).

26

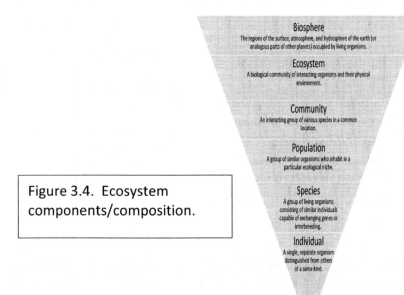

Figure 3.4. Ecosystem components/composition.

Figure 3.5. Illustration of the relationships between complex, nonliner terrestrial and aquatic ecosystems.

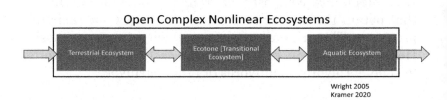

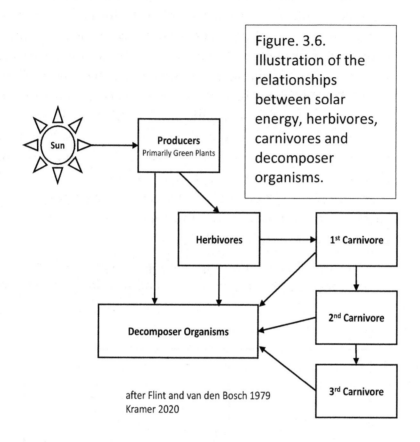

Figure. 3.6. Illustration of the relationships between solar energy, herbivores, carnivores and decomposer organisms.

after Flint and van den Bosch 1979
Kramer 2020

The second trophic group consists of herbivores. These are organisms that feed on plant tissue. Higher level tropic organisms feed on herbivores and are referred to as carnivores (Fig. 3.7). They may also feed on other carnivores as secondary or tertiary consumers. Some species feed on both plant and animal tissue. They are referred to as omnivores. The trophic levels are linked into food chains. Organisms, however, usually interact in complex food webs. Dead organisms undergo the process of decomposition. Through decomposition, carbon dioxide and water are released into the atmosphere. Other chemical elements are released into soil as various system residuals. Two or more organisms interact with each other to form food chains. (Figs. 3.8 & 3.9). Matter and energy are transformed and transported throughout the food chain. In most ecosystems, the food chains expand to form complex food webs (Fig. 3.10). The amount of biomass in an ecosystem is partitioned throughout the various trophic groups. Biomass is greatest in the first trophic group and declines in each subsequent trophic level in accordance with the second law of thermodynamics. (Fig. 3.11 & 3.12). It has been estimated that only about 1.2 percent of the incoming energy from the earth's sun is captured by autotrophs and utilized in the process of photosynthesis (3.13).

Figure. 3.7. Relationships among primary producers, herbivores, carnivores, omnivores and secondary consumers.

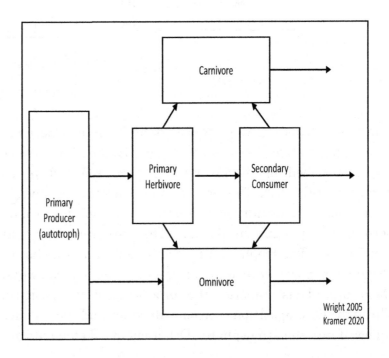

Wright 2005
Kramer 2020

Figure 3.8. Illustrations of three insect food chains.

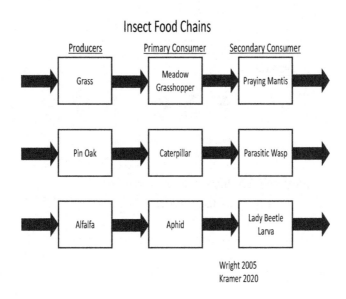

Insect Food Chains

Wright 2005
Kramer 2020

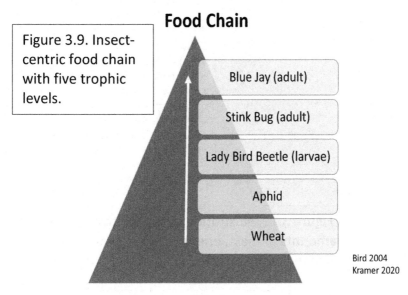

Food Chain

Figure 3.9. Insect-centric food chain with five trophic levels.

Blue Jay (adult)

Stink Bug (adult)

Lady Bird Beetle (larvae)

Aphid

Wheat

Bird 2004
Kramer 2020

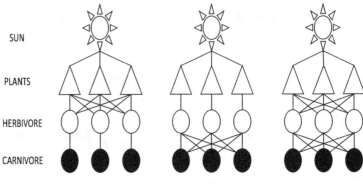

SUN

PLANTS

HERBIVORE

CARNIVORE

Herbivore's Competitors are polyphagous.

Parasites and predators are polyphagous.

Both herbivores and carnivores are polyphagous.

Figure 3.10. Illustration of three types of food webs.

Flint and van den Bosch, 1979
Kramer, 2020

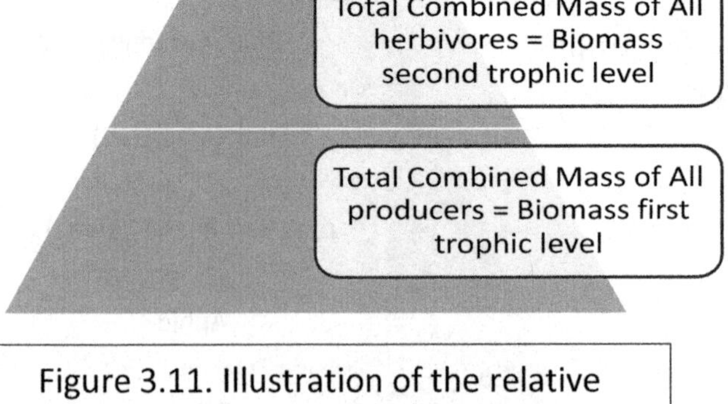

Total Combined Mass of All Carnivores = Biomass third trophic level

Total Combined Mass of All herbivores = Biomass second trophic level

Total Combined Mass of All producers = Biomass first trophic level

Figure 3.11. Illustration of the relative amounts of mass, energy and residuals associated with first, second and third order consumers.

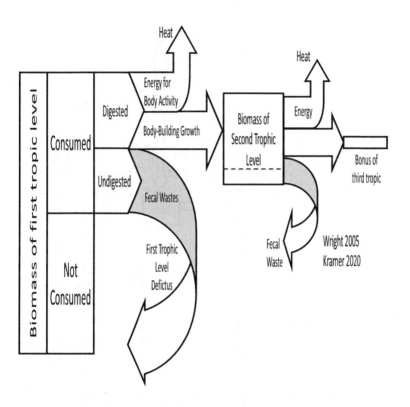

Figure 3.12. Energy flow associated with primary producers, herbivores, carnivores, and decomposers in relation to the amount of solar energy not utilized by primary producers (Kramer, 2020, after Miller 2003).

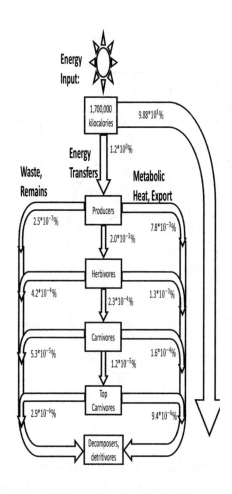

Figure 3.13 Relative amount of energy associated with the five biotic components of an ecosystem.

Sustainability

The term sustainability comes from the Latin word, *Sustiners* (Tinere-to, hold, Sus-up), meaning to maintain or support or endure. In the cases of ecosystem-based and organic agriculture, the farming system should maintain or support the livelihood of humankind on both a short-term and long-term basis. Sustainable systems need to be socially acceptable, naturally compatible and diverse in relation to the flora, fauna and microbial resources they are based on.

Intergenerational equity was included as part of sustainable development in the 1987 report of the Brundtland Commission of the United Nations. Sustainable development, however, is not possible without the synergy of technology, society, economy, and ecology (Bryden, 2002). Murdoch *et al.*, (1992) provided a broad concept of sustainability. They described it as encompassing the viability of localities and communities on which the maintenance of both environment and economic activities depend. Accordingly, sustainable farming focuses on long-term attributes and quality, instead of short-term benefits, high profits and product quantity. Sustainability is based on three biological system principles: 1) the source and sink relationships between inputs and product quality, 2) ecological fitness of species to their ecosystem and 3) population dynamics and biological diversity. The goals of sustainable agriculture provide the option to enrich soil with macro and micro-nutrients, organic matter, enzymes, hormones and vitamins.

The activities of soil flora, fauna and microbes cycle organic residues that protect and enhance soil in a manner not common in conventional agriculture. This farming concept was first mentioned by the

Austrian philosopher, Rudolf Steiner in his 1920 book on *Biodynamic Principles*.

It was followed in 1940 by the writings on compost by Sir Albert Howard. At the same time, Prof. N. R. Dharan and A. C. Acharya were demonstrating composting, with and without mineral additives such as rock phosphate and basic slag.

For the past several decades, there have been many indications that modern fertilizer and seed technologies have reached a point of diminishing returns for increasing crop yield (Flinn and De-Datta, 1984). Moreover, the methodologies of conventional agriculture have resulted in significant losses of top-soil and increases in insect pest problems, plant disease issues, air pollution and unbalanced soil microbial activity. Ecosystem-based and organic agriculture are soil-based systems that mandate that water and nutrients be derived from soil in an effective, holistic, dynamic and environmentally-sound manner. This must be ecologically adapted with optimal socio-economic acceptability.

Relationships

Because of the interspecific and intraspecific interactions among the biotic and abiotic components of agricultural systems, sustainability is not possible from off-farm inputs alone. In the early 1936, Clements, developed the concept of ecological climax or climax communities comprised of plants, animals, fungi and other microorganism residing in or at a steady state of development. This can be referred to as dynamic equilibrium. The process is best illustrated by systems that require minimal physical disturbance. Agriculture, however, requires a certain amount of physical, chemical or biological disturbance. The predictable rhythmic eco-climax of vegetation is almost never present in conventional agriculture. It should, however, be approached in ecosystem-based agriculture on a long-term basis.

Some crop plants have highly specific environmental requirements that are integrated with their physiology. Examples include jute, sugarcane, rice, grapes, cardamom and various other spices. Specific cash crop rotations and cover crops can be used to make necessary environmental adjustments. The theoretical optima, where all the biological functions attain an optimum, is similar to Clements' ecological climax. Deviations from the dynamic equilibrium increase stress and reduce crop productivity. Extending a crop area beyond its limit of natural habitat results in ecological and monetary costs. This is why ecological harmony and soil health are two of the pillars of ecosystem-based agriculture. While soil nutrients may be identical in different ecosystems, their impacts on crops or animals are often not the same (Dash, 1994). An understanding of the relationships among the biotic and abiotic components of a food system is necessary for implementation of ecosystem-based agriculture in a manner designed to achieve sustainability.

Agricultural systems consist of the following four components: 1) abiotic [non-living] parts, 2) autotrophic primary producers [plants], 3) heterotrophic consumers [animals] and 4) decomposers [primarily bacteria and fungi]. The functions of these components are interrelated. Biswas (2014) remarked that the health of individuals (autecology) and communities (synecology) cannot be separated from the health of the ecosystems. For example, healthy soils result in healthy crops that foster the health of animals and people. In addition to being used for food, animals can supply farm energy and soil nutrients (Basu, 2008). The design of ecosystem-based farming systems must facilitate the attainment of ecological balance with local conditions including culture, residue cycling and management of matter and energy. Loss of biodiversity mandates the need for increased energy from off-farm matter as a substitute for essential functions from the system's abiotic and biotic interactions.

Integrated livestock-crop, crop-fish, livestock-fish or crop-livestock-fish systems generate ecosystem synergy. Including system residual cycling as an integral part of the process is beneficial for essential soil, flora, fauna and microbes (Radhamani *et al.* 2003). Combining crops and livestock can have the supplementary/complementary effect of risk reduction, particularly for small farmers in developing nations (Agbonlabor *et al.* 2003; Okigbo,1995). In ecosystem-based agriculture, the residuals (non-food system outputs) from one component become a system input (stimulus) for another. Under optimal conditions, this generates a network of nutrient flow without a need for major external off-farm inputs.

Biological Diversity

Ecosystem-based agriculture is designed to maintain, and in some cases, enhance biological diversity. While at least 1.5 million species of plants have been described, the majority of species of bacteria, fungi and other important ecosystem regulators such as nematodes, flagellates, ciliates and amoebae have yet to be discovered. With the advent of molecular biology, several soil-borne groups of organisms were removed from the Animal Kingdom. This resulted in nematodes becoming the most prominent group of animals on our planet. Nematodes, flagellates, ciliates and amoebae mineralize soil nutrients through their feeding and digestive processes, making the previously sequestered (unavailable) nutrients available for crop growth and development.

Natural ecosystems serve as reservoirs of biological diversity. The greatest number of described species occur in tropical rain forests and coral reefs (Pal, 1982; De, 2010). Among terrestrial ecosystems, tropical rainforests possess about 50% of world's flora and fauna. India, with 329 million hectares of land, has a rich biodiversity and is regarded as a gene bank for many food crops, forest trees, medicinal plants and aromatic plants. Because of accelerated deforestation, urbanization, industrialization, and soil health degradation, many of these species are in danger of extinction (Wilson, 2016). While tropical rain forests contain more than 50% of the world's described species, about 20% of them are facing elimination (Dey and De, 2015). Climate change has a significant impact on species distribution, ecosystem function and crop/livestock production success (Dasgupta *et al.*, 2011 and De, 2010).

The spirit of conservation ethics was present in Indian culture, religion and philosophy during the Vedic era. Beginning in the first century, this was lost as waves of foreign invaders and migrants arrived. The following message of Iso-Upanishad, however, is relevant today: *The universe has been created and nursed by God: man can enjoy the bounties of nature by giving up their greed and not by destroying habitat and biodiversity.* Founded in early Gurukul education system, Seers, Monks and Rishis organized forest heritage centers of meditation (Tapaban Ashrams) to preserve forests and biological diversity. The inherent culture of preserving nature was practiced by Lord Buddha, Mahabir Jaina, and Emperor Asoka, and in modern times by Mahatma Gandhi and the Noble Laureate poet, Rabindranath Tagore. This philosophy remained during colonial times and post-independence days in India. Today, however, it has been lost. As a result, the Indian forest cover has dwindled from 80% to 12% (De, 2014). This change mandates a strong justification for agricultural systems that maintain or enhance the biological diversity necessary to regulate the quality of ecosystems and their external environments. Wilson (2016) proposes that as much as 50% of the earth's surface needs to be conserved for biological diversity maintenance.

System biomass supplies organic matter containing millions of microbes which mineralize soil nutrients and assist in building its essential physical-chemical properties. McNaughton (1979) indicated that dominance of a single species is inversely related to diversity and stability of the ecosystem. Devastation caused by the soybean cyst nematode (*Heterodera glycines*) is a modern-day example. For this reason, the quantification and comparison of species diversity between ecosystems under various climatic conditions becomes important to measure the potential stability of farming systems. The Simpson and Shannon diversity indices are commonly used for this type of assessment. In general, mature and stable communities have high diversity values and unstable communities exhibit low diversity indices, approaching zero (Odum, 1971). Biological diversity varies qualitatively and quantitatively in aquatic, forest, marine, desert, mountain, rural, urban, suburban and agricultural systems. Plants and plant products incorporated into soil can enhance biological diversity and system stability. Some of these have additional value as dietary substitutes or medicinal properties.   In addition to eco-literacy, a general understanding of the history and nature of food systems is important for the development of successful ecosystem-based agriculture.

Chapter 4: Food Systems

Agriculture is a human-managed system of food, feed and fiber production. The laws of nature and science regulate *everything* in the world. Agriculture is no exception. The future of agriculture is dependent on the ability of farmers to be productive in a socioeconomic and environmentally sound manner. Farming is a noble profession. It is a way of livelihood for producing food, feed and fiber. It is the gift of nature, a supreme driving power and force in the universe. Environmental manifestations are what we see or perceive around us as biotic and abiotic forms or images of natural beauty. When harmonious relationships between humans-animals-microbes, flora and fauna, are maintained, ecosystems thrive. Under these conditions, they proceed and progress normally. Imbalance among biota and its environment results in distortion and possible destruction of its ecosystem. Creation and destruction are two facets of the same coin, with a potential endpoint of collapse.

Agriculture began about 10,000 years ago. Since knowledge about ancient agriculture is derived from the archeological record, details about it are limited. Ancient agriculture employed local resources and new innovations. Human lives were based on the natural resources associated with where they lived. Greek, Roman and Vedic literature indicate that people respected the sanctity of natural features. During the passage of time, in the course of societal growth and urbanization, systems of traditional agriculture evolved. Traditional agricultural exists today in many parts of the world. It consists of small to mid-size farms based primarily on utilization of on-farm resources, including human and animal labor. Rapid modernization of agriculture began about 1750 A.D. with the mechanization associated with the western Industrial Revolution. This was enhanced by the discoveries of early agricultural science, such as Liebig's Law of Minimum and an understanding of the causes of infectious diseases. It became known as conventional agriculture and relies heavily on off-farm resources such as fertilizers and pesticides. The post-World War II chemo-technology era resulted in the use of synthetic agricultural chemicals. In many cases, increases in the use of these chemicals had negative impacts on natural resources and human health. Today, it is neither possible nor desirable to return to the agriculture of the past. It is also inappropriate to continue the detrimental and non-sustainable aspects of the current system. There is a distinct need to evolve a new self-regulating, sustainable and equitable system of ecosystem-based food, feed and fiber production of the future.

Ancient Agriculture

Archaeological evidence indicates that the art of growing and preserving plants for food, feed and fiber began about 10,000 years ago in the fertile virgin crescent lands of the East Mediterranean region on both sides of the Nile river. The process included burning forest trees and shrubs, clearing jungles, selecting and planting seeds from native ecosystems and nursing the resulting seedlings. The practice was extended to Iran, Iraq, Palestine, Mesopotamia and Tripoli.

Domestication of cows, goats and horses was a key event. This took place on the eastern sides of the Tigris and Euphrates Rivers and in parts of Syria and Lebanon. The animals contributed to agriculture in a manner that was highly compatible with the civic lives of the people of Mesopotamia and Elam. From 2,350-2,000 B.C., this process took place concomitantly with the Harappa civilization of the Indian sub-continent. The major contributions of Neolithic culture included: animal husbandry, agriculture, building-construction, earthen-crafts, improvement of stone arms, knives, tillage implements and weaving.

During Harappa civilization, oats were brought under cultivation. This was followed by wheat and rice. Rice was first cultivated in the Indus valley low-lands and sides of rivers and lakes. In was grown in China *circa* 3,000 B.C. This indicates that rice was suitable for early traditional agriculture under a range of geographical and environmental conditions. It can be assumed that in these times, the climate consisted of sub-humid to semi-arid conditions. This changed and some areas became rocky and often desert-like. Similar cropping systems were used by nomadic people early in the Stone Age. Ancient cropping systems developed in the steppes of Russia, river sides of the Danube in Europe and lake ridges of Switzerland. These frequently included oats (*Hordeum hexastachium sanctum*) and small grain wheat (*Triticum vulgare-antiquorum*). Two types of jowar (sorghum) were cultivated in Italy and Egypt.

Archaeological evidence indicates that wheat and oats were first grown in Palestine and rice in India *circa* 3,000 B.C. People had become accustomed to village life by 1,000 B.C. in Vedic civilization in the Indian sub-continent. Their main professions were animal husbandry and crop production. Wheat, rice, sugarcane, jowar, black gram, sesame and vegetables were successfully grown in the Indo-Gangetic plains. The farming systems were based on natural resources, use of animal dung and other organic manures. Farmers of this region also grew cotton and used fur and silk for making clothing. High quality silk was imported from China via the Silk Route through the Himalayan Mountains. Bullock-drawn sugarcane crushers were used to make juice from jiggery (cane sugar). Round stone blocks were employed to press seeds for edible oil, mustard and sesame. In addition, the Gabarbandh irrigation system was developed at Mehergarh, Baluchistan *circa* 6,000 B.C. By 1,000 B.C., Atharva-Veda and Parasarasamhita, indicated the importance of the use of animal manures and green manuring (Bisoyi, 2006). In 322-332 B.C., Emperor Ashoka's Finance and Home Minister stressed the use of oil cakes, animal dung and animal urine in the cultivation of crops.

Progressive development in agriculture took place among noble Roman land-lord farmers between 510 and 520 B.C. Their practices were based on rotating pastures with grains, pulses and vegetable crops. Wheat was the most important crop followed by olives, grapes, onions, garlic, mustard, artichoke, corianders, rocket, leeks, celery, basil, mint, rosemary and medicinal plants. The agriculture of the era required a significant amount of labor and slaves were used, especially for caring for cows, oxen, goats and looking after the diary, poultry and bee keeping components of the system. According to the report of Marcus Tarentius Varro, horse manure was suitable for meadow application and poultry and goat manures for grain and other crops. Donkey manure was believed to provide an immediate crop growth response. The Po-Valley in northern Italy was known to be ideally suitable for cultivation of cereals, and the heavy soils of Etruria (Central Italy) for cultivation of wheat. This indicates that early Roman farmers were knowledgeable about ecological aspects of soil for crop productivity. Some Roman farms operated in the northern part of Egypt. These farms utilized the seasonal benefits of the flooding of the Nile River for transporting and replenishing topsoil and nutrients.

Romans were engineers. Their farmers constructed irrigation dams for water conservation. They used mechanical devices for harvesting cereals. Reaping hooks and sheep shears were made of iron. A harvester, called Vallus, was pushed by a mule and developed by the first century A.D. Most farming in Rome was done on large estates. The owners generally lived in towns and the slaves were responsible for the farm labor. The working-class lived in country villages. Roman farms were often irrigated and consisted of poultry, dairy, apiculture, fruit, viticulture and vegetable production. Spain became an early producer of figs, pomegranates, lemons, oranges, lemons and grapes. Vegetables such as squash, eggplant, kidney beans and carrots were important. Cherry production was introduced by the Roman General, Lucullus of Italy. In general, crop yields were poor. There was, however, no shortage of land. Farmers practiced fallowing part of their land to improve soil fertility and conserve water.

In the early period of the Roman Empire, North Africa's rich granaries fed the Romans. Unlike the present times, Roman agriculture was a profession for the Aristocracy. Careful planning went into every detail of owning and maintaining a farm (Cato, 1933). In ancient times, agriculture was not only a necessity, but an activity that supported the elite class of the society as a noble way of life. Around 2,500 B.C. in the Indus valley, at the age of the Mahabharata, the kings of India, such as Maharaja and King Janaka, had thousands of cows, horses, elephants, oxen and vast acreages of cultivated land.

There were also ancient ecologically sound centers of crop cultivation in Africa. These farms were in the Nile River Valley, Niger River Bend, Senegal River Bottom Valley, Lower Congo Basin and Highland of Shaba. Their crops included wheat, barley, millets, sorghum, lentil, olive trees, coffee, banana, rice, yams, rubber, nuts and sesame. Agriculture was practiced on naturally fertile plains and hill slopes.

Excavation at Kuohuqiao indicated that domestication of rice existed in eastern China more than 7,500 years ago and agricultural settlements were present by 500 B.C. Millet was the dominant crop. During the Tang Dynasty (618-907 A.D.), water mills and mold-board plows were used. The ancient agricultural systems of China were natural resource based. With the exception of hybrid rice and other selected crops, emphasis focused on the use of organic inputs. Soil (*Kristi*), water (*Aga*), light and energy (*Teja*), air (*Manut*) and sky (*Born*) were the features recognized as key to successful ancient farming systems (Fig. 4.1).

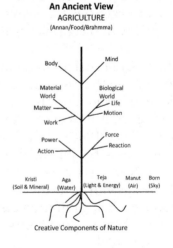

Figure 4.1. Conceptual model of ancient agriculture.

Traditional Agriculture

Traditional agriculture evolved from the farming practices of ancient agriculture. India provides an example of this transition. Traditional agriculture consists mainly of small and medium-size farms using human and animal labor for diverse cropping systems. On-farm resources are used wherever possible. Crops and rotation systems are selected from historical experiences and passed from one generation to the next. Traditional agriculture is commonly practiced today in low-income nations.

Since ancient times, a great diversity of climatic and biological conditions existed on the of Indian sub-continent. These range from the great Himalayan mountains in the north and the Indian Ocean in the south. Natural vegetation and crops are ecologically distributed into fifteen climatic zones. These include highly humid, arid, semiarid, desert, hilly, coastal plains, tropical, sub-tropical and temperate environments. Rice and wheat are the traditional principal crops grown throughout India. Starting from Indus civilization, most crops were grown in the highly fertile soils of Indo-Gangetic belt. These areas were traditionally selected for cultivation of high-quality crops under traditional agriculture practices. Some crops, however, required specific soils and micro-environmental conditions. An example is jute (*Corchorus olitorius capsularis*), a natural fiber with strong threads. It is grown in the eastern humid regions and in Bangladesh. The eastern Himalayan region grows Dehradun/Basmati rice and Darjeeling tea. Black tea is grown in northeast Assam. Traditional farming of apples, apricots, peach, cherry, almond, walnut, and saffron are adapted to the north and western hilly regions. The western dry regions produce jowar, bajra (pearl millet) and livestock; whereas, the central plateau and hills are used for millet, cotton and sunflower. Coconut, casaba and oil seeds have been produced in the western coastal plains.

Traditional agriculture reached North America after the downfall of the Mehergarh civilization near south Turkmenistan and Baluchistan. This most likely was

48

accomplished by bands of ancient skilled tribes moving north and west through Iran, Iraq and the Persian Gulf and eventually crossing the land bridge of the Bering straits (Chakraborty, 1999). They ultimately settled in North, Central and South America, including Nicaragua, Brazil and Ecquador. Native American Indians developed agricultural systems like those of the Mehergarh (Zinn, 2005). They perfected the art of agriculture through natural resource-based farming practices. These included the production of corn, yam and cassava. They developed canal irrigation systems and dams to distribute and store irrigation water. Their talents included making ceramics, weaving baskets and making cloth out of their farm-grown cotton.

In Pennsylvania and upper New York, U.S.A., northeastern Indian tribes known as the Iroquois practiced agriculture. This included traditional systems for growing vegetables, fruit, peanuts, chocolate, tobacco and rubber. Squash was grown as early as 2,300 B.C. and corn by 500 A.D. By the 16th century, the Iroquois and Huron tribes cultivated crops like, maize, potatoes, beans, squash, and sunflower. The Iroquois acquired the skill of de-husking and shelling corn. Their traditional systems of agriculture and types of settlements were almost contemporary compared to the people of far-east Asia, Europe and Africa. They farmed using innovations based on local conditions (Hill and MacRae, 1984).

The history of agriculture suggests that traditional systems of agriculture can be sustained over long periods of time if the natural resource base of the system fits with the crops, animals and management practices used. Use of inappropriate agricultural systems can result in the collapse of entire civilizations, as described by (Montgomery, 2012) in his landmark treatise entitled, *Dirt: The Erosion of Civilization* and Jared Diamond's 2005 book entitled, *Collapse: How Societies Choose to Fail or Succeed.*

Conventional Agriculture

Conventional agriculture evolved over time. As with the transition from ancient to traditional agriculture, it is not possible to provide a date for the beginning of conventional agriculture. Modern conventional agriculture is highly mechanized and involves the use of a significant number of off-farm inputs. It also usually includes off-farm marketing of a larger portion of its products than what is done in traditional farming systems (Fig. 4.2). Following World War II, the use of synthetic chemicals became common. Conventional farms became large and specialized, resulting in a separation of crop and animal production. In the 1960s, high yielding crop cultivars with greater demands for irrigation and fertilizer were introduced. This became known as the Green Revolution. It was implemented in Mexico, India, Thailand Bangladesh, Philippines, China, Japan and other developing countries. Norman Borlaug was awarded a Noble Prize for his innovations related to the Green Revolution.

Before the Green Revolution, India was a "Beg-Bowl" for food. By 2014, the Green Revolution made it possible for India to produce grains close to food sufficiency. To do this, fertilizer consumption increased from below 1,000 tons per year in 1960 to 25 million tons in 2014. Pesticide consumption increased from 200 tons per year in 1950 to 90,000 tons in 1990. In India and other developing nations, these changes resulted in numerous challenging unexpected consequences.

In addition, the technologies of conventional agriculture failed to reach the poor and small land holding farmers of traditional agriculture. This was due to a lack of capital and other reasons such as:

- High cost of production and relatively low profit margins,
- Low quality of produce,
- Lack of rainfall for desired crop yields and high cost of irrigation systems,

- Neglect of dry land farming technologies in India, Africa and Latin America.
- Soil health deterioration, and
- Increased air pollution, water pollution and crop losses due to insects and infectious diseases (Chakraborty, 2014).

Conventional agriculture is significant in China. It accounts for about 10% of the arable land and produces food for 20% of the world's population through the work of about 300 million farmers. Rice is grown in 70% to 75% of the cultivated arable land located south of the Huai river, Zhu Jiang Delta, and Yunnan province. Wheat is grown in the north China plains, and the Wei and Fen river valleys. Tropical fruit is produced on Hainan Island, potatoes in north China and soybeans for tofu and oil seeds in the north and north east China. Green tea is grown on the hill sides of the middle Yangtze Valley. Similarly, India's famous Darjeeling Tea is grown on the hill slopes of Darjeeling and Duars of West Bengal. Silkworm culture is practiced in central and southern China. Much of the crop-geography is based on what ancient and traditional farmers learned about the ecological fitness of specific crops based on local climate and edaphic factors.

Until relatively recently, conventional farmers strongly believed that the next generation of fertilizers, pesticides and genetics would allow for profitability. This, however, is no longer true. It is due, in part, to the development of aggressive populations of weeds, insects and pathogens that cannot be controlled with resistant varieties. Low profit margin is another very important constraint. It is now recognized by many conventional farmers that they must change. Development of an awareness of soil erosion and other aspects of soil health degradation related to biological diversity is a primary catalyst for this change in grower perception. One option is to initiate a new era of nature-loving and ecosystem-based food production systems.

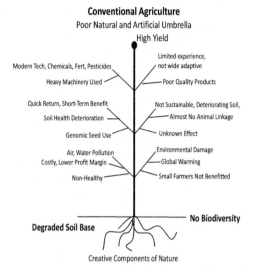

Figure 4.2
Conceptual
model of
conventional
agriculture.

Conventional Agriculture
Poor Natural and Artificial Umbrella
High Yield

Limited experience,
not wide adaptive

Modern Tech, Chemicals, Fert, Pesticides

Heavy Machinery Used — Poor Quality Products

Quick Return, Short-Term Benefit — Not Sustainable, Deteriorating Soil,

Soil Health Deterioration — Almost No Animal Linkage

Genomic Seed Use — Unknown Effect

Air, Water Pollution — Environmental Damage
Costly, Lower Profit Margin — Global Warming

Non-Healthy — Small Farmers Not Benefitted

Degraded Soil Base   **No Biodiversity**

Creative Components of Nature

Ecosystem-Based Agriculture

Ecosystem-based agriculture has been chosen by the authors as the pillar (foundation) of future global food security. This decision was made because of its potential breath of utility and the fact that it is based on the fundamental laws of science and nature. Ecosystem-based agriculture is a holistic, non-linear, open system approach employing the concepts of modern western science and nature-loving aspects of the ancient philosophies of India. It employs decisions that require moral and ethical judgements. Through short-term viability, it assures long-term high-quality environmental sustainability. It also provides positive ecological and humanistic attributes with local and global implications. It is a natural cyclic management system designed to result in a high quality of life for the current generation, while conserving resources in a way that assures a similar or even better quality of life for future generations in all countries of the world. (Aubert, 1972). At maturity, ecosystem-based agriculture becomes a self-regulating system. In the hierarchy of farming systems, the authors propose that organic agriculture is a specialized type of ecosystem-based agriculture.

Agriculture has always been a risky endeavor. Throughout history, innovative farmers worked to reduce the risk of drought, soil health degradation and negative impacts of pests and diseases. They were always searching for ways to improve crop and livestock productivity and quality in a more efficient manner. In more recent times, agricultural scientists have participated in attempts to resolve these issues. These endeavors have resulted in alternatives that are significantly different than the traditional and conventional practices of most farmers. In addition to crop and livestock production, a significant number of alternate systems focus on environmental quality, human health and intergenerational equity. Some of the alternatives are referred to as biodynamic farming, biological agriculture, organic agriculture, natural agriculture, sustainable agriculture and many others. Currently, there is significant confusion about different types of agriculture and their potentials for human quality of life in a sustainable and equitable manner.

Some high-income nations have specific laws and regulations about certain types of agriculture. Low-income nations, however, may only be in the initial stages of planning for their future food, feed and fiber production systems. Unfortunately, when survival is the predominant driving variable in these situations, sustainable and equitable development is not a high priority. The question is often raised about alternate systems and the possibility that adoption would result in decreases in food productivity and security. The answer is no. When properly implemented, these alternate systems can compete with conventional agriculture. Ecosystem-based agriculture has been selected as the pillar of global food security because it combines the best of the agricultural systems into an ecologically-friendly, biologically rich, science-based, socially acceptable and profitable system of philosophically sound practices.

Ecological Harmony.- Agriculture is never practiced as a closed system beyond the influence of natural ecosystem phenomena. It is an open system that takes in matter and energy from external sources. It is biologically nursed through environmental systems composed of clean air, fresh-water, healthy soil, biological diversity and human ingenuity/work. It is an approach to food, feed and fiber production, and other necessities of life through natural on-farm processes used to produce high quality products over a long-term sustainable period. It, therefore, recognizes the whole agronomic and environmental plateau with nutritional and biological components to maintain a healthy, rich soil of diverse microbial activity. The objective is to maintain soil as a place where the multitude of microscopic living organisms, that regulate the processes of life, thrive.

Success in ecosystem-based agriculture depends on properly balancing nature with human-managed crop production goals and practices designed for the specific system. This requires monitoring the biotic and abiotic components of the system. Baseline information about the soil health, pest and disease risk, crop health and status of beneficial organisms is needed to be able to measure change and determine appropriate management actions. Accurate biological and environmental monitoring can be difficult. To do this properly, the owner, farmer manager or farm consultant must have an understanding of the ecology of the farm in addition to its goals and management practices. To achieve this, there are times when the farmer is required to play the roles of scientist and engineer, in addition to that of the production system manager. The theoretical optimum provides a useful approach to this matter. Accordingly, where the different ecological components of nature operate within optimum limits, crops performance is high and can be measured by long-term average productivity. In 2019, Basso et al., referred to this as yield stability. Since nature regulates all complex open non-linear systems, humankind only has limited control over these processes. It is because of this that complex open systems, including agriculture, must be managed and not controlled!

The overall goal of ecosystem-based agriculture is for the farmer to manage the system in a sustainable manner. Since the system is continually changing, the biological and environmental monitoring must be repeated periodically. Information from the monitoring is used in the farmer's decision-making process. The monitoring is then repeated to determine if the correct decisions/actions were made and the system is proceeding in a proper manner. Some components of the system, such as when the monsoon occurs in India, are not controllable in an economic and socially compatible manner. They are non-controlled components of the system and must also be monitored. Even if it was possible to control these phenomena, it is highly likely that they should not be controlled since the consequences of this are often unknown.

The 2015 Global Summit of 195 countries, including India, was held in Le Bourget, France. The delegates agreed to take appropriate measures to limit global warming to below 2 C or even below 1.5 C to combat undesirable climate change. Both India and China are major emitters of carbon and other greenhouse gases that prevent infrared radiation from exiting the troposphere in a normal manner. The ancient Sanskrit literature of India-Rig Veda outlines an environmental philosophy appropriate for ecosystem-based agriculture. Sarbang Khalwadang Brahmma believed that all key components of nature are essential for system optimization, including agriculture. In the Holy Geeta, it is also stated, nothing wrong was created by nature or God. In this context, problems associated with deforestation and over-use of water are created by humankind neglecting the ecological harmony of nature. According to Pancha Bhuta in the Upanishads (2,000 to 3,000 B.C.), nature can be subdivided into five components: Kristi (soil), Aga (water), Teja (energy), Manut (space) and Born (cosmic radiation), as illustrated in Fig. 4.1. In a similar way, sustainability science identifies the five precepts of nature as clean air, fresh-water security, healthy soil, biological diversity and human ingenuity.

In *Ecological Crop Geography* (Klages, 1942), Tensely states that the organic world is composed of physical, mechanical, chemical, biological and anthropic interrelationships. Success in agriculture can only be achieved when Tensely's relationships are close to optimal. This results in a balance between utilization and conservation. Conversion from conventional to ecosystem-based agriculture is a process that begins with development with the goal of reaching a dynamic equilibrium. The time required for this transition varies and is dependent on the overall health of the ecosystem at the beginning of the process. Based on Professor Bird's experiences related to soil microbial information, the process can take as long as seven years (Bird and Smith, 2013). Development or restoration of natural resources, therefore, is a major time-consuming undertaking.

During the Neolithic Age, people, started farming to satisfy their food requirements. This included domestication of wild varieties of cereals such as millets, oats and, wheat etc., in addition to animals. National and regional geography and associated natural biomes/ecosystems dictated the types of agriculture developed and adopted. A thorough understanding of the past provides concepts essential for the development and implementation of a sound and sustainable system of ecosystem-based agriculture. Its philosophy, systems and practices are strongly founded in histories of ancient, traditional and conventional agriculture. **Operating under the pillars of ecosystem-based agriculture is imperative for global food security.**

Soil Health.- Healthy soil is one of the five prerequisites for a reasonably high quality of life for humankind. The significance of soil is well recognized in ancient, traditional and conventional farming systems. In ancient agriculture, the importance of soil was based on individual experiences over a limited number of generations. In traditional agriculture, soil management was based on centuries of experiences passed on from one generation to the next through local experiences or written histories. These included information from events such as the U.S. Dust Bowl that resulted in the formation of the United States Department of Agriculture, Soil Conservation Service. Contemporary soil science, however, evolved primarily in support of conventional agriculture and the commercial fertilizer industry. It was not until the late 1990s that the term soil quality began to appear in the literature. This was followed by the concept of soil health.

In December of 2011, the farmers of the Michigan Potato Industry Commission realized that their soils could no longer decompose corn stalks. They believed this was the result of poor soil health. They requested Professor Bird to investigate the matter. In the summer of 2012, soil samples from 96 Michigan potato fields were sent to the Cornell University, Soil Health Laboratory for analysis. The laboratory uses four physical, four biological and four chemical indicators in their assessment protocol. These include indicators like aggregate stability, nitrogen mineralization and active carbon. Each analysis generates a score ranging from 0 to 100, with a score of the complete analysis of 80 and above considered as a healthy soil. The average score for the 96 Michigan potato fields was 57.7. This indicates that the potato farmers were correct. Their fields had significant soil health problems. Samples from fields with crop rotations three years in length or longer had scores greater than those with a two-year crop rotation. Among the few fields with rotations longer than three years, the highest scores were from those with the greatest crop diversity. One case included seven plant species grown in a five-year rotation system. Five of the plants in this system were cash crops and two were non-cash cover crops grown for soil health enhancement purposes. Subsequently, the topic of soil health has become a topic of considerable discussion in U.S. agriculture. Many farmers realize that their practices of conventional agriculture must change to those of ecosystem-based farming.

Plants obtain most of their nutrients for growth and development from soil. A large portion of the atoms in soil are attached (fixed) to carbon by co-valent bonds and not readily available for uptake by plants. This is often referred to as sequestration. The nutrients must be released from the carbon during the process of mineralization before they are available for uptake by plant roots as ions. Nutrient mineralization is a multistep process. It usually begins with organic matter decomposition by bacteria and fungi. During this feeding process the nutrients are incorporated into the bodies of the bacteria and fungi and are unavailable for uptake by plant roots. While some specialized bacteria mineralize nutrients such as nitrogen on their own, bacteria in general have a low carbon to nitrogen ratio, resulting in a high nitrogen content. In soil, which is often a nitrogen-poor environment, the nitrogen within most bacteria is conserved and sequestered in their bodies. Bacteria either undergo binary fission, are eaten by other soil-borne organisms or enter a resting-stage dormancy before they are of an age to die of natural causes. In general, those that are eaten, are consumed by soil-borne organisms with carbon-nitrogen ratios greater than those of bacteria. This means that the bacteria predators receive more nitrogen than their body requires. As a result, the excess nitrogen is excreted into the soil rhizosphere as inorganic ions that can be taken-up by plant roots. Healthy soils have the biology essential for nutrient mineralization; whereas, soils with poor nutrient mineralization potential are not healthy soils.

The results of the Michigan soil health survey showed that the conventional growers did an excellent job maintaining optimal soil chemistry for high potato yields. Their soil pH, nitrogen, phosphorus, potassium, magnesium, iron, manganese and zinc were at appropriate levels for potato production. The practices that gave these results using commercial fertilizers were based on many years of experience and fertilizer-based soil science education. The soil health problems identified in the survey were associated with the physical and biological attributes of the soil. The physical deficiencies included poor soil water capacity, high soil surface compaction, high subsurface compaction and low soil aggregate stability. The biological deficiencies consisted of low soil organic matter, poor nitrogen mineralization potential, inadequate soil respiration and low active carbon. The tools available for soil health remediation and maintenance relate to the physical, chemical and biological parameters of soil. In ecosystem-based agriculture, special attention must be given to soil biology. Some soil management practices are universal. Others, however, are regional or even site specific.

On a global basis, soil erosion is the most significant issue related to poor soil health. Plato and Aristotle recognized the detrimental influence of accelerated soil erosion through water and wind. Archaeological studies confirm the subsequent decline of Middle East, Greek, Roman, Mesoamerica and other societies was due, in part, to soil health degradation (Montgomery, 2012; Marsh, 1864; Dale and Carter, 1955; Judson, 1968). In these cases, soil erosion and land degradation had taken place for centuries from overuse of high-quality agricultural land and exploitation of marginal land not suitable for agricultural purposes. Soil erosion is caused primarily from the use of marginal slopping lands, excessive cultivation and failure to keep the soil covered with appropriate vegetation.

Soil organic matter (soil carbon) plays a pivotal role in soil health, plant nutrition, and agricultural crop productivity. It is composed of living organisms, including plant root tissue, organic matter in the process of decomposition and ancient stable carbon-based matter. Soil organic matter facilitates water infiltration and retention of essential ions, in addition to preventing soil erosion. The ions are bound ionically to the soil mineral and organic matter. The ability of a soil and its microbes to release these ions from organic matter and the atoms from the bodies of the microbes is referred to as active carbon. Active carbon is a key soil health indicator. It is an assessment of the microbial mineralization potential of the soil that makes the inorganic chemical elements required for plant growth and development available for plants. These beneficial attributes of active carbon are not present in commercial fertilizers.

A significant number of natural resource and ecosystem-based friendly soil management practices have been developed. These include agroforestry, social forestry, grass culture, watershed management, biomass generation through cropping systems both under rain fed and irrigated conditions, residue cycling, and soil conservation (Nitant *et.al.*, 1998). These systems are based on:

1. Utilization of local organic resources, integrated nutrients and input management from different types of ecosystem-based farming such as duckery, poultry, fishery, fruit and vegetable systems.

2. Use of bio-intensive pest management (BIPM) and integrated pest management (IPM).

3. Composting or vermicomposting with or without fly ash, rock phosphate, gypsum or sulfur dust.

4. Green manuring, legumes or fodder inter cropping, bio-regulators and growth promoting microbes-EM (effective microorgaisms).

5. System planning in accordance with biological diversity and local resources, market orientation, education and socio-economic persuasion.

Soil temperature significantly impacts how ecosystem-based agriculture soil is managed. In 2018, Bird and Basso reported a relationship in Michigan between soil health, potato crop productivity and thermo-stability. It is known. that at high temperatures in tropical arid and semiarid soil, organic matter degrades rapidly to 0.3-0.5% in six months; whereas, in temperate regions it can remain at 3-4% soil organic matter (Chakraborty and Sen, 1967).

Soil organic matter can be maintained by regular applications of manure or organic residues. In Italy, Bulgari *et al.*, 2015, found bio-stimulants highly promising. Bio-stimulants are extracts from plants such as neem, basil or sea weeds. They contain a wide range of organic molecules which activate plant metabolism. Other bio-active compounds include vitamins, amino acids, chitin, chitosan and poly-oligosaccharides. In 2006, FAO reported that about 15 million tons of sea weeds are used in agriculture annually as nutrient supplements and bio-stimulants. The modes of action of most bio-stimulants in plants are unknown and under investigation (Guinan *et al.*, 2013). They may involve complex physiological functions between plants and the environment. They may even be part of a plant's defense systems for reducing biotic and abiotic stress and acting as phytoalexins or anti-oxidents (Klichestein, 2004; Bartwal *et al.*, 2013). Characterization of bio-stimulants, therefore, should be done on the basis of plant responses, since their biochemistry and physiology related to the environment is unknown. For example, a product like Radifarm for tomato, may do well in a temperate climate, but not have a positive impact under tropical conditions. Positive responses were obtained for broccoli, lettuce, pepper and potato by spraying Sea sol, Actiwave, Radifarm, and Sea sol, respectively (Haider*, et al.*, 2012). Bio-stimulants appear to be suitable for use in ecosystem-based agriculture.

Soil amendments are often used for maintaining, remediating or enhancing soil health. They include super-compost, phospho-humin, manure, cow dung. incubated rock phosphate manure, vermicompost, vermi-composted fly ash, phospho-bacterin-compost with Azotobacter or *Tricoderma virdi*. Soil amendments are appropriate for use in ecosystem-based and organic farming systems. In 1973, Professor Chakraborty studied the nature of competition between weeds and rice under dry land conditions. He found that bacteria play active roles in soil, such as, improving water stable aggregate formation, minimizing soil erosion, enhancing water infiltration and providing a supply of nutrients for mineralization. They also ameliorate soil acidity/alkalinity and provide a habitat for other soil-borne organisms. Without soil organic matter and its associate microbes, soil is inert matter and its only use in agriculture is as a substance to physically support plant tissues and a place to apply fertilizers and retain water (Russell, 1965).

Long-term experiments at Rothamsted, U.K. and Coimbatore, India indicate that roots and crop residues can provide ten tons of organic matter per hectare without the use of animal manure or commercial fertilizer. Mulching with organic refuse is also beneficial in preserving soil moisture and preventing soil erosion. Utilization of biological-based fertilizers has not been widely adopted even though it is well-known that symbiotic bacteria such *Rhizobium* spp. and asymbiotic *Azotobacter, Clostridium, Beijerinkia* and *Azospirillum* fix atmospheric nitrogen and make it available to plants. An alfalfa-rhizobium association under tropical conditions can fix up to 500-600 kilograms of nitrogen per hectare per year. Similarly, in wet land rice, about 79 kilograms of nitrogen fixation per year has been achieved with micro-organism inoculum. A specific example with rice is *Azolla-Anabaena* biomass providing up to 40 kilograms of nitrogen in 30 days. When the mushroom fungus, *Pleurotus sojar-caju*, is grown on commercial basis, the bedding residues can be used as a source of nitrogen. These technologies can be useful in the ecosystem-based agriculture transition processes. In addition to organic matter, rock phosphate/basic slag, gypsum and sulfur dust can be used to resolve specific soil deficiencies (Peters,1977).

The technology of using effective microorganisms (EM) with or without bokashi (fermented straw or organic matter) was developed by Teruo Higa of Ryukyus University of Japan in 1980. In 1999, Sangakkara provided guidelines for its use in organic farming at the International Nature Farming Research Centre in Japan (Xu, 2013). EM is a mixture of microbial cultures including the photosynthetic bacterium *Rhodopseudomonas* spp., yeast (*Saccharomyces* spp.) and lactic acid bacteria (*Lactobacillus* spp.). EM is sprayed on the crop and bokashi is spread in between rows as a mulch.

Bokashi is made from crop residues, animal dung, urine, green weeds, rice husk, oilcakes, press mud, bagasse (dry pulpy residual of sugar cane after juice extraction), kitchen wastes and fish or bone meals. Moisture must be kept at between 30-40% at 25-35C for 7-8 days, or until fermentation is completed. Bokashi can: 1) enhance photosynthetic activity, 2) promote plant growth, 3) increase nutrient mineralization, 4) provide resistance against insects and foliar diseases, 5) suppress soil-borne pathogens, and 6) improve the overall physical-chemical and biological conditions of soil. EM-Bokashi technology, however, depends on the availability of local resources, prevailing soil temperature and the farm manager's understanding of the process.

Composting is often a key practice for successful ecosystem-based agriculture. It was first demonstrated in North India in the early 1900s by the British agronomist, Albert Howard (Biswas, 2014). The process was developed and popularized by Acharya as the Bangalore method of composting. Organo-phospho-compost was promoted by Prof. Dhar (1930-1940) for improvement of soil fertility and productivity. The use of panchagavya, a mixture of cow dung, urine, and milk, curd and ghee can be used in organic farming to accelerate plant vigor. This product is known as T-7 and includes leaves and basil extract (Prabhu et al., 2010).

The use of bio-stimulants or phyto-chemicals is becoming increasingly important in ecosystem-based agriculture. In the U.S.A., 25% of the 2020 Spring Cushman Creek Newsletter was about bio-stimulants. Bio-stimulants have been reported to reduce nitrate accumulation in leafy vegetables, increase chlorophyll and carotenoids, and promote growth by stimulating root activity and antioxidants. They, also, increase the efficacy of mineral nutrients and reduce nutrient leaching. Bio-stimulants can, also, provide resistance to drought, freezing and salt tolerance. Further, as reported by Gaje-Wolska et al. (2013), the use of phytochemicals can increase photosynthetic activity and improve tolerance to infectious diseases caused by bacterial, fungal and virus pathogens. In India, in 1989, the senior author reported observing decreases in incidence of insects and diseases when rock phosphated-compost was used.

Soil health renovation and maintenance is a global imperative. For success, soil health renovation and maintenance mandates appropriate soil biology and a good working knowledge about how soil works. Reports at the 6th International Conference on Soil Conservation (ISCO) describe the extent of damage of natural resources in different countries. The problem of acute soil erosion has been documented in Africa, Latin America, India, China, Thailand, Vietnam, Far East Arabia and Russia, particularly on the small farms covering more than 20 percent of the world's population. Professor Bird has observed severe soil health degradation in Central Asia in Tajikistan, Uzbekistan, Kyrgyzstan and Kazakhstan. Poor soil health has a direct negative impact on the quality of life of small land holding farm families. Soil health degradation is also an important issue in high income nations in Europe and North America. The Dust Bowl crises in the U.S.A. and former U.S.S.R, are extreme examples. It is known that about 50% of the rich topsoil of the mid-west U.S.A. has been eroded away. Farming practices, that use heavy equipment can leave soil bare, leading to the current soil erosion/health crisis (Morgan, 1986).

Success of ecosystem-based agriculture is dependent on the diversity of soil-borne organisms, including bacteria, fungi, bacterial-feeding nematodes, ciliates, flagellates and amoeba. Populations of these organisms decrease when marginal land is used for monoculture in the absence of cover crops or non-till practices (Leser, 1980). Glasstetter and Leser, (1987) showed that earthworm activity was related to soil losses in Central Europe. They found that soil erosion involves both biological and cultural factors. Earthworms played an active role in soil aggregate formation, indicating a need for a holistic research. In temperate latitudes, average soil contains 325 grams of living microorganisms, mainly bacteria and fungi and 50 grams of soil-borne animals per square meter. Sixty percent of the animal biomass belongs to earthworms (Dunger,1983), which have positive impacts on soil aggregate stability. Research has demonstrated that water stable aggregates prevent soil surface sealing crusting and run-off, thereby reducing soil erosion. Cropping systems influence earthworm populations. Their activity is increased under grassland and decreases when arable cropping is practiced, resulting in erosion and reduced soil carbon (Tisdall, 1985).

Excessive physical disturbance of soil results in poor soil health. In Venezuela, Paez and Rodriguez (1992) reported that agricultural practices are frequently done under sub-humid to semi-arid conditions using heavy machinery on marginal land with high rain, and erodible soils. This results in extensive soil health degradation. In a CHA-alfisol, with a 4% slope, soil loss was 31 tons per hectare per year. In a MAR-mollisol with a 15% slope, sheet and rill erosion was 573 tons of soil loss per hectare per year. Cultivation of corn or sorghum provided little protection; whereas, grass culture reduced erosion to 3.7 tons per hectare per year in a CHA-soil. The same was observed in Ethiopia where large scale farms using heavy equipment resulted in soil erosion of 27 to 94 tons per hectare per year (Grunder, 1985). Hurni (1983) reported that at the Didessa State Farm, 50% of the land has been impacted by erosion, making about 2,070 hectares unfit for cultivation. Corn yield was reduced from 3.0 to 2.3 tons per hectare with mono-cropping on marginal land (Tafesse, 1992). Use of contour grass strips and crop rotation with maize-sorghum-sunflower-groundnut-soybean and minimum tillage demonstrated that soil erosion in Ethiopia could be reduced to 15 to 10 tons per hectare per year, compared to the previous 94 tons per hectare per year.

While some soil health issues are global in nature, many are regional or site specific. The following seven geographical case studies illustrate various soil health issues and potential short-term and long-term solutions:

1. The main food crop in Nigerian agriculture is cassava (*Manihot esculanta*). It is grown on more than seven million hectares. This represents 53% of global production. It is grown as an intercrop with maize, cowpea, sorghum, okra, and melon (Okigbo, 1981; Okigbo 1978). In 1976, Aina *et al.* reported that mono-cropping of cassava resulted in greater water runoff than with maize and cassava intercropping. Mono cropped cassava resulted in soil losses of 110 tons per hectare per year; whereas, the intercropping system with maize reduced soil loss to 69 tons per hectare per year. In addition, the intercropping improved soil bulk density, water infiltration rate, water retention capacity and earthworm activity. The results included increased crop canopy development, biomass generation, soil organic matter, soil biology and root volume; all of which are beneficial for protection from soil erosion. This, however, cannot be generalized due to local conditions of soil types and availability of sufficient moisture for biomass decomposition.

2. In Mexico, fertile flat land represents only 14% of the farm-land base. Demand for this scarce resource is increasing. Use of marginal land becomes unavoidable, resulting in as much as 70% of the farms suffering from soil erosion (Geissert and Rossignol, 1987 and Maass, 1992). Maize is the main crop. It is grown as a monoculture with poor irrigation capability and low yields. Conversion of virgin tropical forest land to corn production results in poor soil health. Research, however, indicates that on hilly slopes, use of maize stover mulch is an effective conservation strategy for reducing soil loss and increasing crop productivity. In a maize mono cropping system, soil loss was 100 tons per acre per year, while the mulch system resulted in a loss of only five tons per hectare per year (Maass, 1992). The impact of mulching was observed early in the crop growing period, at a time when the crop canopy

had not filled-in the rows and the soil was directly exposed to rain and wind. While wheat was introduced into Mexico in 1520 A.D. by Spaniards, conventional wheat production was achieved through the use the of fertilizers and dwarf gene varieties of winter wheat associated with the Green Revolution. This, however, was limited to the 14% of the best irrigated land. This system is very different than the wide scale corn cultivation in hilly, mostly unirrigated erosion prone areas of Mexico.

3. India has been an agricultural country since ancient times. Seventy percent of its population of 1.3 billion are directly involved with agriculture. With 2.4% or 329 million hectares of the world's land, Indian agriculture cultivates 173 million hectares in support of 15% of the world's population. This is done through zone cultivation patterns throughout the different states of India. India's major food diet consists of rice and wheat, and to a lesser extent jower (sorghum), bajra (pearl millet) and maize. Due to the high demand for food security, commercial fertilizers and pesticides are used in conventional agricultural systems on 60% of India's irrigated flat lands. This is practiced for rice, wheat, sorghum, maize, sugarcane, potato and vegetable production. Under rain-fed or partially irrigated conditions in hilly states like Uttrakhand, Kashmir, Meghalaya, Nagaland, Manipur, Mijaram and in dry-land arid zones, small growers depend mostly on organic manures, oil cakes and ash with or without commercial fertilizers or pesticides. This can be considered as a hybrid traditional/conventional system.

About 60% of the farmers in India are small/marginal growers. The majority were not able to afford the input costs associated with the Green

Revolution. In addition, soil health degradation has become a serious threat to Indian agriculture. In 1974, it was recorded that 80 to 100 million hectares of land had been impacted negatively by accelerated soil erosion. The extent of this depends on the slope of the land, with soil loss being greatest on roadside slopes, irrigation canal sides, river sides, coastal areas, undulating land mass of arid, semi-arid, and lateritic red soils and mountain terraced land during the rainy season. Large scale soil and associated nutrient loss also takes place when cultivated/puddled rice land is annually inundated by flood water. This results in an average annual loss of about 127 kilograms of nitrogen, 57 kilograms of phosphorus, 1,056 kilograms of potassium and 6,093 kilograms of calcium per hectare which is equivalent to 5.4-8.4 million tons of these elements per year.

Chakraborty (2014) reported that about 57% of India's land resources have undergone various types of soil health degradation. These include acidity, alkalinity and salinity. About 3 to 5 billion tons of topsoil is eroded every year by water and wind, resulting a direct loss of plant nutrients (NPK) of 8-10 million tons annually. This loss is primarily due to: 1) exploitation of fragile marginal land for cultivation, 2) urbanization and industrialization, 3) faulty soil-water management in cultivation, 4) over-exploitation and removal of ground water, 5) encroachment of forest undulating land for cultivation with or without terracing, 6) intensive cultivation of cereals without cover cropping, inter cropping, green manuring, mixed cropping, or in situ disposal of green matter, 7) deforestation and allowing silting of river beds and 8) permitting run-off water without taking any scientific measures for water conservation.

Ecosystem-based agriculture provides low cost soil conservation and overall soil health maintenance. Hallsworth (1987), an eminent soil scientist, indicated that the most modern effective soil conservation is achieved by keeping the land green with a vegetative cover and reducing the gradient of slope of surface soil over which run off water takes place. Cover cropping, strip-cropping and mixed cropping are age-old practices in India and many Asian countries. In areas of low rainfall, a vegetative soil cover can be difficult to sustain. Work at IARI (Delhi) indicated that berseem clover planted between rows of sugarcane can improve soil structure, and resist water run-off and soil erosion; whereas, cultivation of maize, potato and wheat resulted in deteriorating soil health. In 1971, Professor Chakraborty found that the drought resistant pasture legume, *Atylosia scaraboides*, can form a thick cover on soil within two months and resist soil erosion, in addition to increasing soil organic carbon. In the hilly soils of Dehradun, with an 8% slope, giant Star grass was more effective than maize-wheat or contour cultivation of maize/cowpea/wheat resulting in about 10 tons per hectare per year less soil loss. *Stylosanthes* spp., a perennial pasture, green manure and efficient fodder legume crop contains about 18% protein, 29% fiber, 44% soluble carbohydrates, 4% Ca and 6% P, besides contributing 240 kg N/ha to the topsoil. Under medium to low land conditions, *Sesbania rostrate*, raised as a pasture legume and green manured resulted in 5.6 tons of dry biomass equivalent to 84 kg N/ha from leaves and 28 kg N/ha from stems. *Crotalaria juncea* can supply 10 tons of dry biomass/ha and nearly 196 kg N/ha in eight weeks.

Sikkim agriculture in the Himalayan regions is ecosystem-based. The system closely resembles organic agriculture. It employs strip cropping, cover cropping, mixed cropping, crop rotation, mulching, agro-forestry, and utilizes farm-yard manure. The farmers adopted indigenous management practices to reduce soil erosion and water runoff. These include terrace cultivation, soil-water retention walls and water ways. About 23% of the cultivated land is under terraces used for rice, mostly by conventional shallow tillage. The remaining 77% of the land is under the age-old practice of slopping terraces (Mishra and Rai, 2013). Alley cropping with fodder legumes, *Ficus, Atrocarpus Bauhinia and Leucaena* species are grown as strip crops on the edges across the slopes, along with vegetative crop barriers. These systems conserve soil, water and nutrients very efficiently. The innovations are evidence that regenerative and resourceful technologies can provide environmental and economic benefits to farmers, communities and nations (Pretty, 1995).

4. In Thailand, alley cropping with *Leucaena leucocephala*, drilled in rows and cut at 1.5 m in height every two months, resulted in 120 tons/ha of green matter/ha. In 1-3-year-old tea plantations in Assam, India, a popular practice is to grow a dense legume cover between rows. In 1992, Professor Chakraborty reported that this system results in a soil cover within a short period of time by cutting the top six to ten inches of the plants and using it as a living mulch to protect top-soil and also supply nutrients. In the absence of animal manure or compost, green biomass can be generated on-farm and used to restore soil health. Indian rice farmers have four months of rain and an eight-month dry season. In wet soils, farmers incorporate the remaining six to eight inches of rice stubble after harvest into the soil. This is done with manure, with or without *Sesbania aculeate*, as additional organic matter to stabilize soil structure and prevent erosion (Bhattacharya, 2011).

5. In 1998, Forge reported that degradation of agricultural land is a problem in Canada. This was due to wind erosion, surface water run-off, salinity, acidity, soil compaction and loss of organic matter. The practice of growing corn and soybean in a legume rotation has been replaced by an era of growing a corn/wheat rotation. This resulted in increased soil degradation in the prairie areas of Canada and the U.S.A. Following the droughts of 1977 and 1980,

farmers began to use soil conservation measures, such as minimum tillage, direct seeding and forages/grasses on marginal lands. These innovations include wind breaks and permanent grass barriers. Value soil loss cost in Canada in 2000 was estimated to be $2 billion (Agric. Canada).

6. In Australia in 2004, Loch stated that the health of soil is likely to be deteriorated when agriculture is practiced on low productive lands without following appropriate ecological methods. In this context, crop rotation, fallowing and practices designed to increase soil organic matter are recommended. These, however, must be site specific based on temperature and rainfall conditions. In his work at Darling Downs, Loch remarked that practices are often not appropriate for the specific ecosystem. Darling Downs is an area of about 1.8 million hectares. It is where Australia agriculture began in the 1840s. When the high demand for food occurred after World War II, use of conventional agriculture increased, resulting in soil health degradation. This included soil erosion and intensive gulley formation. Today, reduced tillage, changes in cropping frequency and retention of crop residues are common. On steep slopes where rainfall is only 250-400 mm annually, there has been only limited success in prevention/stabilization of soil erosion.

7. Many experiments in the U.S.A. and Europe indicate that soil erosion rates associated with conventional agricultural are greater than rates of soil production. With no-till agricultural, there was a 75% reduction in soil-erosion in corn fields in Indiana (Montgomery, 2012; Johnson and Moldenhauer, 1979). In Ohio, Harold *et al.* (1970) found a more than 10-fold decrease in soil loss from no-till compared to conventional agriculture. In tobacco production, a 90% reduction in soil erosion was associated with no-till farming compared to conventional practices

(Wood and Worsham, 1986). The extent of reduction of soil loss by no-till practices compared to conventionally ploughed soil depends on various factors, like, type of soil, land slope, amount and frequency of rainfall and other local, cultural and soil biology. No-till agriculture adoption has increased as a cost-effective alternate to conventional tillage. In 2000, 16% of the cultivated area on U.S.A. farms used no-till. (Derspch, 2001). It has been Pofessor Bird's experience that some growers swear by no-till farming methods and others swear at no-till technologies. There is usually a learning curve associated with successful no-till farming. No-till can involve increases in herbicide use. When a local Illinois corn/soybean conventional production system was compared with 1) ridge tillage with pesticides, manure and green manure and 2) crop rotation with legumes, ridge-till and legume-based Illinois-tillage, corn yielded 104-121% and 92% of the conventional system, respectively. Since the no-till learning curve is steep, on a global basis, farmers are in a dilemma as to which technology to use. No-till benefits include soil erosion reduction, soil moisture enhancement, greater organic matter and increases in soil biological diversity,

Pest and Disease Management.- Humankind has always had to deal with the detrimental impacts of pests and infectious diseases of plants and animals. Throughout history there were years of abundant harvests, followed by years of low crop yields. Famine caused by high population densities of locust feeding on wheat plants or late blight of potato are examples. Innovative pest management practices were used as early as 2,500 B.C. A serious locust plague, however, took place in eastern Africa in 2020 A.D.

Pest management can be divided into four eras: 1) pre-synthetic pesticide era, 2) synthetic pesticide era, 3) era of integrated pest management and 4) era of genetically modified organisms. Sulfur was used as a pesticide by the Sumerians as early as 2,500 B.C. In some places, ancient pest management became part of religion and superstition. This led to unsuccessful laws to outlaw pests. Roman-engineers built rat-proof granaries. By 600 A.D., mercury and arsenic were used in China for pest control. Homer reported the use of fire for locust control. The first report of biological pest control was from China in 250 A.D. In association with the plague in Europe, women wore flea trap necklaces. The Industrial Revolution and mechanization of agriculture fostered numerous early discoveries about the biology of pests. The first books devoted to pest management were published in the 1800s. Koch's postulates provided a new understanding of the causes of infectious diseases (Koch and Loeffler, 1876). The pre-synthetic pesticide era included the use of many inorganic pesticides including sulfur, copper, arsenic, lead and mercury. Successful mosquito control allowed for the much-delayed construction of the Panama Canal.

The synthetic pesticide era began immediately after World War II. It was initiated by the discovery that DDT had insecticidal properties. This was followed by extensive use of chemical-warfare nerve poisons, including organo-phosphates and organo-carbamates, for pest management. Pests rapidly developed resistance to these pesticides. Rachael Carson's land-mark book, *Silent Spring*, identified the serious unexpected consequences of the synthetic pesticide era. These included: 1) development of pest resistance to pesticides, 2) chemical contamination of the environment, 3) acute and chronic human health risks, 4) harm to non-target beneficial organisms, 5) pesticide induced evolution of new key pests and 6) pest population density resurgence. These events resulted in the development of a new ecosystem-based management system that became known as integrated pest management (IPM).

IPM was defined by Professor Bird in 1977 as the *development, use and evaluation of pest control strategies and tactics that result in favorable socio-economic and environmental consequences.* Two years later, in an environmental message to the U.S.A. Congress, President Carter, a former conventional peanut farmer, defined IPM as *a systems approach to reduce pest damage to tolerable levels through a variety of techniques, including predators and parasites, genetically resistant hosts, natural environmental modifications and when necessary and appropriate, chemical pesticides.*

IPM is based in Professor Ray Smith's (University of California-Berkeley) action threshold concept (Stern *et al.* 1959). It involves extensive monitoring (scouting) of both the biotic and abiotic components of the ecosystem prior to selecting and implementing a pest management practice. There are four IPM strategies. These include 1) pest avoidance or exclusion, 2) pest containment or eradication, 3) pest control (population reduction) and 4) do nothing. Under pest control, there are six available actions including, 1) physical, 2) biological, 3) chemical, 4) cultural, 5) genetic and 6) regulatory tactics. Sound pest management practices are essential for successful ecosystem-based agricultural

On a global basis, the over dependence on pesticides and narrow genetic crop-base has fostered   serious insect pest and infectious disease problems resulting in significant crop yield losses. The situation is particularly serious in tropical countries. Crop pests and infectious diseases are major issues in India, Japan, China, Philippines, Bangladesh, and Uganda. This is true especially on small farms where insect pests in rice, wheat and maize are abundant. These are mostly monocropping systems that make integrated pest management difficult. Increased pesticide use will not solve the problem.

Use of cultural methods for pest control has been emphasized (Upadhyay *et al.*, 1984). Although rice is a principal food, in developing countries such as India, Bangladesh, Thailand, Japan, China, Vietnam, Myanmar,

Ethiopia, Uganda and India, yields are low and special pest management programs are required. There are, however, long-standing constraints to successful pest management programs. These include: 1) poor bio-physical conditions of soil, 2) socio-economic conditions of small and marginal farmers and 3) technological and institutional issues (Mahapatra, 1990). In addition, rice productivity is impacted by rainfall, drought, flood, climate changes and poor drainage. The lack of adequate irrigation facilities and poor maintenance of soil organic matter is common. Most organic crop refuses and dung are used as fuel. Animal husbandry is absent or poor, and no green manuring or crop rotation is practiced. These factors impact other crops in addition to rice and detract from the potential benefits of high-quality pest management. Professor Bird's experience in the U.S.A. indicates that initially, it is best to work with a few key early adopter farmers. When these individuals are successful, most of them will automatically serve as educators for the early-majority farmers. This indicates that successful IPM is both a community-based and overall ecosystem-based process.

In 1993, it was reported that under puddle conditions of rice, insects and other pests were only a minor problem. This is not true for the tropical rice belt, where the practices of ecosystem-based agriculture are very limited. An exception is scented rice production. In wet-land rice in Tamilnadu, Kerala, India, duck foraging resulted in a 97% reduction of red and white rice weed seeds in winter flooded fields. Ducks and geese also feed on the seeds of barnyard grass, smart weeds, beggar ticks, crab grass, panicum and some other insect pest species.

Overall, there are relatively few pest or infectious disease resistant varieties commercially available. Some are GMO (genetically modified organism) varieties. These are not popular in various regions. Many farmers are not well trained in IPM. This results in the overuse of pest resistance cultivars and development of highly aggressive pest/pathogen populations. This is especially serious with the soybean cyst nematode (SCN) in the U.S.A. As a result, a multistate coalition was developed to provide the information necessary to alleviate development of highly aggressive SCN populations. One unique aspect of the program is that MorganMeyer, a major private sector marketing firm, was hired to make SCN management information available to all soybean growers in the U.S.A. This information is electronically available at www.thescncoalition.com. In 2020, the Coalition was awarded the National Agriculture Marketing Association Grand Champion prize for public relations.

While use of GMO crops is increasing, it is limited to specific crops like, corn, cotton, squash, soybean, potato, sugar beet, tomato and sweet pea. GMO crops have been banned or have limited use in many countries. In nations like U.S.A., Canada, Germany, Italy, Switzerland, Sweden and Australia, the use of bio-pesticides and bio-stimulants is becoming popular among ecosystem-based farmers. For organic farmers in Afro Asian countries, these products are costly and not readily available. These farmers prefer to use mechanical and indigenous procedures. Light traps, yellow sticky traps, bird perches, hand picking of larvae, changing sowing dates and barriers are employed in fruit, vegetable, tea, coffee, other horticultural crops, scented rice and medicinal plants. These are used for crops grown especially for export purposes. Tamilnadu Agritech Portal (TNAU, India) demonstrated pest control for leaf eating caterpillars and borers with *Andrographis paniculate,* neem kernel extracts and garlic-chili-ginger extracts. Cow dung extract is recommended for control of pumpkin beetles and pod sucking bugs. For tomato wilt, Fusarium wilt in peppers, yellow mosaic virus, Alternaria leaf spot and vegetable fruit rot; spraying with 10% cow urine and application of 40 kg neem cake was effective. Although these indigenous practices are cumbersome, they can be an option for small growers.

Complete freedom from insects and diseases in agriculture, involving plants and animals, is neither possible nor desirable. The cost-benefit ratio would be prohibitable and it would also result in the destruction of many beneficial insects and animals in the soil-ecosystem, resulting in negative unexpected consequences. Plant protection in ecosystem-based agriculture can be achieved two ways: 1) by direct pest population reduction, including chemicals such as Pyrethrum, or Neem extract or by 2) indirect management practices such as crop rotation, cover cropping, inter cropping, strip cropping, cropping barriers, use of bio stimulants, bio agents, phytohormones, and seaweed application. In conventional agriculture, the use of pesticides at recommended doses can result in long-lasting damage to soil, aquatic ecosystems and human health. It is also well-known that increased use of pesticides can result in pest population resurgence and new pest issues. In addition to growing pest and pathogen resistant varieties, cultural, mechanical and traditional methods can provide excellent results when implemented in an appropriate manner. A few of the options include zero tillage, minimum tillage, deep ploughing, weathering or sunning of ploughed land, crop rotation, cover cropping, crop barriers, water logging, use of light traps, burning crop residues, biological control agents, sanitation and drying harvested grain. In ancient agriculture, hand picking was used to remove Papilio caterpillars from young citrus trees. Fallowing crop land for one season is often effective to control insects and diseases. Weeds need to be removed since they can be alternate hosts for insect pests and infectious disease pathogens. Application of organic manures in various forms such as farmyard manure, animal dung manure, compost, super compost, vermicompost, phospho-mineral-compost with basic slag, gypsum, sulfur dust, poultry manure and green manuring can improve soil health, crop vigor and crop tolerance to insects and diseases. The microbes associated with soil organic matter play important roles in suppressing crop pathogens, including phytopathogenic nematodes. Professor Bird's favorite hypotheses states that, if a site has a soil-borne disease issue, it does not have heathy soil.

Organic materials contain bio-active compounds such as vitamins, amino acids, chitin, chitosans and poly-oligosaccharides. In 2006, FAO reported that bio-stimulates, can reduce soil-borne insect pests. In addition, some pest or pathogen resistant crop varieties are approved by certifying agencies for use in organic farming systems. As reported from studies in Odisha, India during the wet season, growing the hybrid high-yielding rice varieties RH.10428 and RH.10422 lowered the incidences of gall midges, brown plant hoppers, leaf rollers and stem borers. These appear to be options for use in IPM associated with ecosystem-based agriculture. The availability of specific resistant varieties frequently changes. This can be a positive or negative factor. It is a negative if a specific variety is not available or a positive if it has been replaced by an even better variety.

Although there are acute and chronic human health risks associated with synthetic pesticides, these chemicals can be used in ecosystem-based agriculture on an emergency basis as long as they are applied in a manner that is not harmful to natural resources or human health. Synthetic pesticides, however, are prohibited in organic agriculture. Use of other synthetic inputs for plant protection purposes, including pheromones are consistent with the philosophy of ecosystem-based agriculture. These, however, should be integrated with cultural and mechanical pest control practices and indirect plant health nutritional enhancement methods. Experiments conducted in a rice-wheat (no-till)-green cropping system with vermi-compost, farmyard manure, BGA, Azolla, and Azotobacter in IARI, India, during 2003-2006 increased rice yield about 115%. The research indicated that there were no serious insects or disease issues associated with these practices and yields were similar to those of conventional agriculture using commercial fertilizers at recommended rates.

Research indicates that that plants suffering from poor nutrition are often highly vulnerable to insects and infectious diseases. Bio-stimulants can be used to increase plant vigor in a way that protects crops from insect pests and infectious diseases. This can be a low-cost alternative with no human health or environmental risk. Plant extracts and bio-stimulants contain a wide range of organic molecules that can activate plant metabolic processes in short periods of time. They can increase chlorophyll content in leaves, reduce nitrate accumulation in leafy vegetables, stimulate root growth and enhance antioxidant potential. In some cases, they have been shown to enhance soil nutrient uptake. The global market for bio-stimulants is projected to increase by 12% annually (Bulgari *et al.*, 2015). Other scientists have shown that bio-stimulants can enhance plant growth through drought resistance, salt tolerance and improve tolerance to bacterial, fungus and virus diseases (Norrie and Keathley, 2006; Gaje-Wolska *et al.* 2013). Actiwave, Benifit, Viva, IPA Extract, Megafol, Seasol and Radifarm are commercially available bio-stimulants. They are appropriate for use in ecosystem-based agriculture and have been approved for use in organic agriculture by the International Organic Certification Standards. Some caution, however, is necessary since they have not been tested under a wide variety of environmental conditions and their modes of action are not well known (Guinan, *et al.* 2013).

In 2006, FAO reported that about 15 million tons of sea weeds are used in agriculture globally as nutrient supplements and bio-stimulants. They can be used as soil conditioners and plant stimulators resulting in increased yield of potato through application of sea-weed extract, PRIMO. In 2010, at Rajasthan, India, Prabhu *et al.* reported that Panchyagavya-T7 (a mixture of 2% Panchagavya-dung, urine, milk, curd and ghee plus 0.2% HA plus 2% Maringa leaf extract; Basil-*Ocimum sanctum*) kept plants healthy and productive.

The Center for Indian Knowledge Systems on Organic Farming in Tamilnadu, Chennai, developed improved traditional practices to control leaf-eating caterpillars, shoot and stem fruit borers, and hairy caterpillars through application of 3-5% Seriyangai, a mixture using *Andrographis paniculate* and 5% Sida Spinosa with *Arivalmani poondu*. The shoot systems are cut into pieces, boiled in earthen pots mixed with soap water and diluted before being applied as a spray. A garlic-pepper-ginger extract was used to control aphids, green plant hoppers, mealy bugs, and white flies. Application of fresh cow dung extract (1 kg dung plus 10 liters of water or 10% cow urine or neem kernels extract) controlled beetles, pod-sucking bugs, tomato wilt, Fusarium wilt of pepper, Cercospora leaf spot, and yellow mosaic virus disease. Hot water treatment or fumigation of sugarcane sets before planting was effective for controlling red rot disease. This is perceived by growers as being more effective than red rot resistant varieties. These examples indicate that various traditional and low-cost practices have potential for small-scale ecosystem-based agriculture. The case for large-scale conventional systems, however, is questionable.

Landscape ecology is an important aspect of pest/infectious disease prevention and management. In Germany in 2006, Beanchi *et al.* reported that human living environment expansion results in declines in biological diversity and the functioning of natural pest control. The exchange of natural enemies between crop and non-crop habitats is likely to increase in diversified landscapes. Loss of natural ecosystems results in deterioration of the biodiversity related to successful natural pest control (Ives *et. al.*, 2000; Gurr, et *al.*, 2003). In a diversified mosaic agricultural landscape, natural enemies are likely to have year-round sources of food and shelter (Dunning, *et al.*,1992). This indicates that in large areas of India and Bangladesh, where crops are grown continuously throughout the year, such as rice-rice-rice, rice-wheat-green gram or jute-rice-potato rotations; natural enemies are endangered due to frequent disturbances of crop habitats. Some biological control organisms take shelter in distant areas or adjacent ecosystems. The structure of both cropping systems and adjacent ecosystems is therefore important for effective management of insects and infectious diseases.

Biological control of pests is defined as the use of living organisms, broadly called natural enemies, to reduce and maintain pest population densities below their damage threshold. Biological control agents include: 1) predators, such as birds, amphibians, reptiles, fish, and mammals; 2) parasitoids, such as wasps that deposit their eggs in the pest and 3) pathogens that cause infectious diseases of the pest. Important insect predators include lady bird beetles, ground beetles, flower bugs, spiders, and mites. There are four groups of parasitoids. These include: 1. Ichneumonid wasps (prey mainly on caterpillars of butterflies, and moths), 2. Braconid wasps (attack wide range of insects such as green flies, cabbage white caterpillars), 3. Chalcid wasps (parasitize eggs/larvae of green flies, white flies, cabbage caterpillars, strawberry Tortrix moth), and 4) Tachinid flies (attack wide range of insects such as caterpillars, larvae and beetles). *Gonatocerus ashmeadi*, a Hymenoptera sp., was introduced in 2006, in French Polynesia (Hoddle, *et al.*, 2006). It controlled glassy winged sharpshooter. The fungal spores germinate on the larvae, resulting in mortality. For biological control to be successful, it is necessary to 1) conserve natural enemies, 2) not use any pesticides or use in a very restricted manner and 3) culture and release certain new natural enemies after thoroughly studying the ecology of the system. Accurate identification of the host and potential biological control agent is essential before undertaking any biological control practice.

Several species of entomopathogenic nematodes are available for use in ecosystem-based agriculture. They can be purchased or cultured on-farm. For biological control with nematodes to be successful, it is imperative to have the correct nematode species that is specific for the insect species to be controlled. In addition, it is essential to apply it in a proper manner. Nematodes are susceptible to light and desiccation. This mandates that they not be applied during sunlight or under dry conditions. Ground covers and mulch can be used to enhance control through decreasing risk to exposure to light and desiccation. Insect control with commercially available entomopathogenic nematodes can be expensive and often not suitable for marginal traditional farms. Entomopathogenic nematodes are appropriate for use on progressive ecosystem-based, organic and high crop-value conventional farms (Bird, *et al.*, 1990; Bugg and Pickett,1998). With appropriate training, it is possible to successfully culture entomopathogenic nematodes on a small scale on ecosystem-based farms.

Research in Michigan and other locations has demonstrated the value of bio-fumigation for management of specific soil-borne pathogens. Cover crops like mustards and radishes are grown to green pod stage and immediately incorporated into the soil for bio-fumigation. The chemicals released during the decomposition process are converted into isothiocyanates that have fumigation properties. The procedure has been used to control a broad range of phytopathogenic nematodes. In some cases, more than one bio-fumigation cycle is necessary to obtain results similar to those of commercial soil fumigants. Bio-fumigation is appropriate for ecosystem-based and organic agriculture. Commercial soil fumigants harm soil biological diversity and are not suitable for use in ecosystem-based and organic crop production systems (Bird, Wernette and Lott, 2012).

Parasitic fungi can be used to control thrips, white flies, aphids, caterpillars, weevils, grasshoppers, Colorado potato beetle and mealy bugs. They produce toxins in the host-insect body, resulting in insect mortality. If the fungus survives and produces spores, it can persist throughout a growing season. Parasitic fungi for insect control are sold as spores. High doses are frequently used, and the cost can be high. In some cases, application can have a negative impact on non-target beneficial insects. Brown spot of rice

84

caused by *Helminthosporium oryzae* is a key disease of rice in India and other countries, particularly in rice-wheat cropping systems. It has been reported that seed treatment with *Tricoderma viridi* at 4 gm/kg of seeds plus a foliar spray with *Pseudomonas fluorescens* at 10 g/litre resulted in 34% less disease in the presence of *Helminthosporium oryzae*. Neem oil applied at 3% as a seed treatment and neem extract used as a foliar spray (4gm/kg seed plus 10 g/liter) were also effective in controlling this pathogen.

The most commonly used bacterium for insect control is *Bacillus thuringiensis*. This organism is popularly known as BT. The bacterium produces an insecticidal protein that provides effective control of many insect pests. In general, it has little effect on non-target beneficial natural enemies. BT is widely used for control of specific Lepidoptera, Coleoptera and Diptera species. Its toxin genes are incorporated into GMO crop cultivars like corn and brinjal (eggplant).

Biological control agents may be general or highly specific. Some of commercially available insect pathogenic viruses are specific to a single type of insect. Gemstar for corn earworm and tobacco budworm and Spod-XLC for beet army worm control are examples. The advantage of utilizing parasitic viruses for biological control is that they are obligate chemical messengers that replicate within their specific insect host. They provide sustainable, effective and safe pest control.

Naturally occurring biological pesticides are appropriate for use in conventional, ecosystem-based and organic systems. While there are differences among national pesticide registrations, these may include *B. thuringiencis* (bacterial toxin), Pyrethrum (Chrysanthemum extract), Neem (extracts), Rotenone (root extract) and compost teas (liquified compost). In general, use of these materials in the late 20th Century was low (Jones, 1998). Compost tea is biologically active. It contains beneficial microorganisms. and must be prepared under controlled conditions to be an effective product. While certain inorganic pesticides are approved for use in organic agriculture, if they are not properly used, they can have detrimental impacts on the essential biology of ecosystem-based and organic agriculture. Examples of inorganic pesticides include Bordeaux mixture, copper sulphate, copper hydroxide, and sodium bicarbonate.

Pheromones are complex natural chemicals that insects use to communicate with each other for mating and food resource identification. They can be used to disrupt mating or for direct mass trapping of adults. Numerous types of traps, including sticky boards, have been developed for use with pheromones in accordance with the fundaments of chemical ecology. (Gut *et al.*, 2019).

Sharma *et al.* (2014) examined different ways to use algae in crop production. These included seed treatment, foliar application and soil drenching. They found multiple benefits, including enhanced root growth, increased photosynthetic activity, higher crop yield, cold tolerance, drought tolerance and salt tolerance. Algal extracts can induce resistance to fungal bacterial and viral pathogens. According to FAO, commercial algal products for agriculture are available in numerous countries, including China, Indonesia, Philippines, Korea and Japan.

Organic Agriculture

Nature and History of Organic Agriculture.- Organic agriculture is a special type of ecosystem-based food, feed, fiber and livestock production. It is based on healthy soil. Use of synthetic fertilizers, synthetic pesticides or genetically modified organisms (GMOs) are prohibited. An organic label is required for food to be sold as certified organic. This requires that an independent organic certification organization inspect the farm to assure that organic production and processing practices are being used in accordance with a set of organic standards. As a management system, organic practices are designed to promote biologically diverse regenerative soils that produce healthy, nutritious crops and livestock. Certified organic produce and livestock are marketed at premium prices. National and international organic standards are used to describe the procedures required for organic certification. On a global basis, the demand for organic food has increased significantly in recent years.

In the U.S.A., the United States Department of Agriculture, Agricultural Marketing Service is responsible for overseeing and enforcing the National Organic Program. Internationally, CODEX (Codex Alimentarius Commission) defines organic farming as .....*a holistic production-management system which promotes and enhances agro-ecosystems, health, including biodiversity, biological cycles and soil biological activity.* According to IFOAM (International Federation of Organic Movements), *organic agriculture is a production system that sustains the health of soil, eco-systems and people. It relies on ecological processes, biodiversity and cycles adapted to local conditions.*

Organic agriculture on small and medium-size farms combines traditions, innovations and science to benefit the shared environment and promote fair relationships and a good quality of life for all involved. It fosters the use of mechanical, cultural and biological methods such as crop rotation, green manuring, residual cycling, biological mineralization of soil nutrients, organic mulching, use of compost, vermicompost, composting with mineral additives and biological pest control to facilitate a healthy and productive agriculture. Thus, it is a holistic, traditional, but modern approach to agriculture that integrates the specific practices of ancient, traditional, ecosystem-based and conventional farming.

One popular maxim is that when soil health deteriorates, so does human quality of life. This has taken place many times throughout history on farms managed under ancient, traditional and conventional practices. In 1950, Dr. William Albrecht identified the basic principles of measuring the quality of soil health using physio-chemical and biological methods. In India and Sweden, as early as 1930, Prof. N. R. Dhar demonstrated the benefits of mineral-phosphate compost in cropping systems without addition of fertilizers. In many respects, organic agriculture is not new. It is a process based on the best practices of traditional and ecosystem-based agriculture that focus on soil health.

The concept of organic farming was proposed in the 1920s by the Austrian philosopher Rudolf Steiner, and again in 1940 by Albert Howard of the British Association of Organic Farming. In general, there are two types of organic farming. Certified organic farms grow crops for premium prices at domestic or export markets. Non-certified organic farms are usually small and grow for local markets where they have established customer relationships about the quality of their produce and production practices. Non-certified organic produce is also grown in countries that do not have certification standard laws. At the International Conference on Organic Agriculture and Food Security in 2007, it was agreed in principle that organic agriculture has the potential to feed the world provided it is practiced in the correct way. This mandates good soil health and use of organic pest and disease control procedures.

Changing from conventional or traditional to certified organic agriculture requires a transition period. Three years is the usual transition period time required for certification.

Professor Bird's research results associated with converting a conventional corn-soybean system with poor soil health to a certified organic apple orchard showed that in this case, it took seven years for the soil biology to reach a state of dynamic equilibrium. Initially, crop productivity is lower under organic than conventional management. It has been demonstrated, however, that by year seven, productivity under organic practices can equal to or exceed that of conventional production.

Research has demonstrated that under environmental stress conditions in rain fed and arid zones, organic agriculture out-performs conventional methods. One example is the Rodale Institute Long-Term Research Trial that compares organic and conventional practices. In market-marginalized areas, poor organic growers can increase production using local organic resources without relying on external inputs. When low yields occur, they can be compensated with premium prices or government incentive programs.

In a research trial in India, organic productivity was 9% lower, but had a 22% higher net profit, compared to conventional production. This was due to premium prices of 30-40% and reduction of input costs by 12%. When off-farm organic inputs were purchased from distant markets and price incentives were not available, organic farming was not economically viable. In a study in Uttaranchal, Madhya Pradesh, and Tamilnadu, India, comparing conventional and organic farming on 40 farms, it was observed that rice yields were lower in organic farming. Net marginal income, however, did not differ significantly. Input costs were lower in organic farming system than that in the conventional one (Table 4.1) In addition, a model used in the study showed that large scale conversion to an organic system in Europe and North America should not have a major impact on food security. It was concluded that organic farming is a viable option for improving food security of some small farms.

According to the United Nations projection, global human population is expected to increase from 7.5 billion to 9.7 billion in 2050. This means that food production will have to increase. Organic farming will play a key role on the environment, soil health, animal welfare and rural development associated with this change. On its own, however, it will not be able to address the vast and perplexing issue of solving global hunger and food security. The global position of organic farming is encouraging. For the most part, organic agriculture involves farmer and consumer choices, with an emphasis on soil health, without focusing on profit maximization. Corporate organic, however, attempts to maximize profit and often ignores soil health.

Organic farming is practiced in 170 nations on a total of 43.7 million hectares or 1% of agricultural land. According to a 2015 FiBL-IFOAM survey, most of this acreage is in Europe and North America. Among developing countries, China has 2.3 million hectares and India 1.2 million hectares of organic crop land. Products like organic soybeans, rice, tea, fruit, meat, coffee, vegetables, flowers, rice, tea, medicinal plants and fish are exported mainly to Europe and North American.

India annually grows 585,970 tons of organic produce worth $301 million for export. Organic aquaculture exists in China, Bangladesh and Thailand. In 2012, the International Federation of Organic Agriculture estimated that the Global demand for organic products is increasing at approximately $5 billion per year. While consumer demand for organic products has increased, the percent of agricultural land under organic production remains small (Tables 4.2 & 4.3).

Table 4.1. Results of a comparative study of conventional and organic farming in India with an organic premium of 20% of gross margin (Paramaiyan *et al.*, 2009).

Parameter	Conventional	Organic
Tamilnadu Region		
Rice yield (kg/ha)	4,270	3,400
Input cost (Indian Rs/ha)	3,967	2,682
Net margin (Ind.Rs/ha)	25,900	21,835
Uttarakhand Region		
Crops production (kg/ha/yr)	2,845	3,250
Home consumption (kg/ha/yr)	2,046	2,168
Total net margin (Rs/ha/yr)	14,500	15,200
Madhya Pradesh region		
Irrigated cotton yield (kg/ha)	1,694	1,260
Rain fed cotton yield Kg/ha)	1,187	1,044
Net margin without premium Irrigated cotton (Ind. Rs/ha)	28,974	25,234
Net margin without premium Rain-fed cotton (Ind. Rs/ha)	18,900	19,541
Net margin no premium (Ind.Rs/ha) Irrigated cotton	28,974	30,730
Net margin with premium (Ind.Rs/ha)- Rain fed cotton	18,900	24,134

Table 4.2. Percent of agricultural land in organic production in 2013 (Ciccarese and Silli, 2016).

Australia	2.8	Argentina	1.7
China	0.4	India	0.5
U.S.A.	0.5	Italy	9.0
Uruguay	6.1	Germany	4.8
UK	3.8		

The 2000 Indian Organic Farming Plan focused on popularizing organic farming in localized zones. It was designed to stimulate biological cycling, conservation, watershed management and technology with local organic resources. The National Program of Organic Production (NPOP) was initiated in 2001 at the Gaziabad Institute in Delhi. The program had a potential for 9.2 million hectares of organic farming (including forests). The NPOP began the processes of accreditation and farmer certification. It is important to note, however, that in India and on global basis, the number of non-certified farmers using organic practices is significantly greater than the number of certified organic farmers.

Table 4.3. Percentage of agricultural land in Europe (Ciccarese and Silli, 2016)

Spain	14.3	Italy	10.4
Germany	9.3	France	9.2
Poland	5.9	United Kingdom	5.3
Austria	4.8	Turkey	4.7
Czech Republic	4.4	Sweden	4.3

No-till farming has been especially difficult in organic agriculture. The benefits of no-till, however, are highly compatible with organic agriculture. The challenges of no-till farming in organic agriculture is related to: 1) weed management, 2) manure and compost incorporation, 3) green manuring and 4) elimination of perennial legumes or winter annual cover crop residues before planting; all of which normally require at least minimum tillage (Teasdale, 2016). This frequently results in growers selecting a minimum tillage system and not no-till. Jeff Moyer of the Rodale Institute, however, developed a roller/crimper that is used in a cover-crop system with cereal rye for preparing a weed-free seed bed for organic agriculture.

Professor Pimentel of Cornell University indicated that conservation of soil, water and other natural resources including high soil organic matter, high soil nitrogen and low fossil energy inputs resulted in organic crop yields that are similar to conventional systems. He found that this is especially true in drought years. Conventional corn farmers in the U.S.A., however, used synthetic fertilizers and pesticides, resulting in negative impacts on public health and the environment. As a result of herbicide use, health care costs were estimated to be about $12 billion per year (Pimentel et al., 1993; Pimentel, 2005).

Results from the Rodale Institute multi-decade corn-soybean trial comparing organic and conventional practices are very encouraging in favor of the organic system. This project was designed by the late Dr. Richard Harwood. In most years, yields were similar. In drought years with soybeans, the yield was lower in organic farming system; however, organic corn without premiums was 25% more profitable than conventional corn. Corn was grown 60% of the time in the corn/soybean conventional rotation; whereas, only 33% of time in the organic system. The extra yield and profit from wheat grown in the organic system compensated for the difference in the rotation system. A conceptual model of ecosystem-based and organic focuses on optimal yields, enterprise sustainability and soil health (Fig. 4.4).

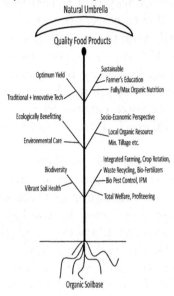

Figure 4.4 Conceptual model of ecosystem-based organic agriculture.

Global Organics.- As consumers become more conscious and concerned about the quality, safety and nutrition of their food, there is an increased interest in organic agriculture. Certified and non-certified organic acreage has increased significantly. In 2017, the global organic market was estimated to be more than $100 billion. The majority of consumers of organically growing foods are in North America and Europe. Organic agriculture has been increasing in Australia, Argentina and China. The highest number of organic farmers, however, is in India with 340,000, followed by Uganda, and Mexico with 180,000 and 130,000, respectively. China, Bangladesh and Thailand have large acreages devoted to organic farming. In 2006, the ten nations with the greatest percentage of their farming acreage under organic practices range from 7% to 29% farmed organically (Table 4.5). In a 2010 study of 12 nations, the percent of land farmed organically ranged from 0.3% to 12.8% (Table 4.7). When summarized by continent, the range was 1% to 28% (Table 4.6). Long-term demonstrations like the Rodale project and those of early-adopter local organic farmers resulted in increases in the global status of organic agriculture (Fig 4.3).

Table 4.5. Land farmed organically in ten nations in 2006 (FIBL, 2008).

Liechtenstein	29%	Austria	13%
Switzerland	12%	Estonia	9%
Greece	8%	Portugal	7%
Sweden	7%	Latvia	7%
Italy	9%	Timor Lester	7%

Table 4.6. Percent organic farms by continent (FIBL, 1998)

Oceania	1%	Asia	13%	Africa	24%
N. America	2%	Latin Amer.	3.2%	Europe	28%

Table 4.7. Status of organic farming in twelve nations/regions in 2010 (Willar, 2010).

Country	Agric. Land	Country	Agric. Land
Australia	12.8%	Argentina	1.7%
Africa	3.0%	Asia	9.0%
China	0.4%	India	0.3%
Italy	9.0%	Germany	4.8%
N. America	8.0%	Uruguay	6.1%
U.K.	3.8%	U.S.A.	0.5%

Figure 4.3. Organic agriculture status.

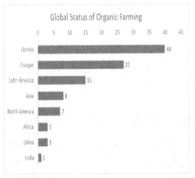

Organic Farming area of total cultivated land

Highest organic wild products: China and India
Global demand increases @ $5.0 billion/Year
Maximum demand: Europe and North America

In 2008, Australia had the most land in organic production, with 12.3 million hectares (Table 4.8). This was followed by Argentina and China with 2.2 and 2.0 million hectares, respectively. The amount of land in the ten nations with the greatest amount of wild organic ranged from 7.5 million hectares in Finland to 0.5 million hectares in Sudan (Table 4.9).

In 2012, IFOM indicated that ecosystem-based/organic farming is practiced in 160 countries. Ten of the nations have 70% or 26.3 million hectares of the land used for organic production. Australia, Germany, and Denmark are major organic producers with Australia having the greatest amount of organic grazing pasture. Organic farming is practiced on 37.5 million hectares. About 0.7 million farm families practice organic or ecosystem-based biological agriculture. Europe is responsible for 30% of the world's organic agriculture. This is followed by Latin America with 18%. Consumers demand for organics is greatest in North America and Europe. The global demand for organic products is increasing at more than $5 billion per year and is estimated to have been $100 billion in 2017, compared to $18 billion in 2000.

Table 4.8. Ten nations with the most land in organic production in 2005 (FIBL, 2008).

Country	Area
Australia	12.3 million hectares
Argentina	2.2 million hectares
China	2.0 million hectares
U.S.A.	1.6 million hectares
Italy	1.1 million hectares
Uruguay	0.9 million hectares
Spain	0.9 million hectares
Brazil	0.9 million hectares
Germany	0.8 million hectares
U.K.	0.6 million hectares

Table 4.9. Ten countries/regions with largest wild organic (Willar, 2010.

Nation/Region	Area
Finland	7.5 million hectares
Zambia	7.2 million hectares
Brazil	5.6 million hectares
Azerbaijan	3.2 million hectares
China	2.9 million hectares
India	2.4 million hectares
Serbia	1.1 million hectares
Bolivia	1.0 million hectares
Argentina	0.6 million hectares
Sudan	0.5 million hectares

INDIA: The current goal is to popularize improved traditional, ecosystem-based and organic farming systems. The Sevagram Declaration was initiated in 1994 as part of the Gandhian Village Swaraj and the Chipco Movement in 1973. In 2000, a 10th Plan-2000 thrust for agricultural development focused on popularizing organic farming using specific technologies such as local resources, recycling and conservation. The National Program of Organic Production (NPOP) was initiated in 2001 and accreditation of organic certification agencies began in 2001. After 2001, the National Institute of Organic farming at Gaziabad, Delhi initiated organic farming research on site-specific cropping systems, crop nutrition and soil fertility management technologies. As a result, India converted more than 9.2 million hectares to organically managed land by 2009. This included 8.0 million hectares of organic wild forest production. By 2012, India became the largest organic cotton growing country, producing 77,000 metric tons. India also has the highest organic forest wild harvest. There are about 340 thousand certified organic farms in India, accounting for 0.3% of the total agricultural land. Most of the organic produce are exported and local consumption is negligible.

It was determined that the organic farming system in India was relatively unorganized. Most of the farmers were not aware of its long-term benefits to soil health. This resulted in a government incentive subsidy in cash or kind, for organic inputs for small and marginal farmers that had not adopted organic practices. There are no organic markets at the state or district levels and no market orientation or premium prices for organic producers. In general, consumers are not interested in purchasing organic produce at comparative high prices. Because of the demand to achieve food security and zero-hunger, farmers are more interested in high yielding conventional technologies under irrigation. Recently, however, the NPOF and NCOF are trying to increase farmer interest in organic production through training, demonstrations, quality control, certification standards, organic market development and financial supports for biological fertilizers and other organic inputs. In West Bengal and other states, bio-village recommendations have begun as recommended by the Swaminathan Group in the late 1980s (Biswas, 2015). Special emphasis on organic farming practices in dry land areas, hilly states of Meghalaya, Nagaland, Manipur Arunanachal Pradesh, Mizoram, Kashmir and Uttarakhand were initiated because the feasibility of growing fruit and vegetables, tea, coffee and other horticultural crops under ecosystem-based or organic technologies.

As stated by the Prime Minister of India, Mr. Narendra Modi in 2016, the Himalayan state of Sikkim became India's first fully organic area. This was achieved by converting about 75,000 hectares of agricultural land to ecosystem-based farming. In most parts, the entry and sale of chemical fertilizers was banned. The major organic crops in Sikkim are cardamom, ginger, turmeric, vegetables and flowers such as, cymbidium, rose, gerbera, anthurium and mandarins. Organic buckwheat, millet, maize and rice are also grown. Under organic farming, Sikkim produces nearly 80,000 million tons of farm products every year by implementing organic packages and practices that assure good markets for the farmers.

Sikkim started an organic market complex represented by farmer groups to sell their products at premium prices. The products are exported to locations in mainland, India. The population of Sikkim is about 610,000 with more than 100 villages adopting bio-village tilth in 2009. Each village was assigned to grow a single crop using ecosystem-based farming practices. This assures quality management and marketing. The total geographical area of Sikkim is 729,900 hectares. Farming is done on about 10% of the land. The area is divided into many small plots, mostly in mountain terraces and valleys. The remaining land is under forest, permanent pasture or other culturable land.

Sikkim has multiple agro-climatic zones. These include tropical, sub-tropical, temperate, sub-alpine, and alpine zones. Most of the agriculture is done in the tropical, sub-tropical and temperate zones. More than 26,000 hectares are certified organic and about 28,500 hectares are in transition for certification. This represents slightly more than 64,000 small farms. The certified crops use cow dung manure, compost; vermicompost, biofertilizers, rock phosphate, EM solutions, and bio-stimulants, or neem extract. Due to unavailability of reliable irrigation, farmers practice mostly rain fed and integrated farming systems with animals and horticultural crops. Since Sikkim is situated at the base of the Himalayan forest range, the ecology and its biological diversity are ideal for organic farming. The soil contains 2% to 7% organic matter. The executive body for organic and ecosystem-based agriculture is the Sikkim Organic Mission (SOM). This is under the chairmanship of the Minister of FSAD & HCCD. Policy protocols are implemented by the Sikkim Government through SOM. ICS development is achieved through service providers and certification through APDA accreditation agencies. Incentives for organic growers include on-farm input production, training programs and market linkage.

CHINA: China is the second largest country in the world in terms of certified organic farmland. This labor-intensive system includes an extensive amount of wild organic farm systems. Northern China's long and cold winters and cool and dry summers create a favorable environment for ecosystem-based and organic farming. The major organic crops are soybean, vegetables, rice and tea. These are exported to Japan, Europe and North America. Organic rice and tea are grown in ecosystems with good biological insect and disease management resources. There is no national subsidy or other support for organic growers. The perspective for China's organic food production, however, is positive. Its volume is expected to increase 30% to 50% percent to obtain 1% of total agricultural production. The biggest challenge facing the organic farming industry in China is the need to improve the domestic market. Consumers are uninformed about organic products and their benefits. They are skeptical about organic certification. An additional problem is the poor understanding of organic regulations and weakness of enforcement of organic standards. A survey indicates that farmers complain about problems associated with organic pest and disease management. In ecosystem-based and organic agriculture, farmers use biological pesticides approved by the China National Accreditation Services and Accreditation Administration of China (Scoones and Elsaesser, 2008).

A large portion of China's 1.3 billion hectares of agricultural land has poor soil health. Soil degradation is estimated to be 78%, 72% and 90% for China's crop land, forest and grasslands, respectively. Ecosystem degradation and pollution are common in rural areas. Government, however, plays an important role in promoting organic agriculture. Organic production provides new opportunities for poor farmers (Willar and Lernoud, 2014). Chen and Scott (2014) observed that China's path to organic agriculture is based on export markets through contract farming. Domestic consumers and markets, however, have emerged gradually.

The diversification of farm ownership rights and profits has played an important role since 2000. Farmer co-operatives serve as alternatives for purchasing inputs and marketing produce.

AUSTRALIA: Australia has the largest organic farming industry in the world with 22 million hectares. It produces a variety of products including beef, fruit, vegetables and poultry. These are raised without synthetic fertilizers, synthetic pesticides or genetically modified seeds. Revenue was projected to grow at 14% in 2015-2016. Organic meat is raised in a range land pastoral system housing cattle, sheep, hogs and turkeys. It is a highly capital-intensive industry. The early Australian economy was based on sheep (wool and meat), cotton, wheat, fish, and beef industries. Recently, pork, seaweed, oranges, apples, bananas, chestnuts, potatoes, carrots, and tomatoes have become important crops. Australian farmers are world leaders in dry land farming, natural resource management and sustainable agriculture. Australian consumers are showing increased interest in food quality and the environmental benefits related to organic products. Australia is a major agricultural producer and exporter of produce raised mostly under natural range land and dry land management systems. About 325,300 people are employed in its agroforestry, fish, hog and poultry industries. In 2015, this amounted to 12% of the nation's GDP. Although most organic produce is exported, Australian farmers supply 93% of Australia's food. In Australia, the organic industry is economically viable.

UGANDA: In Uganda, organic or ecosystem-based agriculture is the best option for small scale farmers to be economically viable. With a population of about 10 million and a favorable environment, Uganda is the food basket of their region. Most products are exported at high prices. Very few local people have the purchasing power to benefit from organic foods. There are no large-scale organic farms.

The small-scale organic farmers are registered and trained. Their farms are inspected throughout the growing season. Produce is certified organic. Farms in transition from conventional to organic are registered and monitored. The organic market, however, is limited. Farmers do not receive premium prices for their produce until they have completed the certification process.

Uganda has taken various unique steps in transforming conventional agriculture to organic farming. It involves unique coordination of all stake holders, including the Minister, Export Promotion Board, National Bureau of Standards, Coffee Board, Cotton Development Organizations, Private Institutions, Farmers Associations and Export Companies. This is achieved under a common umbrella, called, NOGAMU. Uganda organic products include cotton, coffee, sesame, dried fruit, avocadoes, jack fruit, vanilla, cocoa, fish, shea-butter, nuts and honey.

Uganda has more than 400,000 internationally certified organic farmers. This ranks first in Africa and second globally. The total area under organic production is nearly 350,000 hectares, representing more than 2% of the total agricultural land. The organic industry grosses about $37 million per year. The country's fertilizer consumption is less than 1.0 kg/ha, perhaps the lowest in the world. Further, it is important to note that organic and ecosystem-based agriculture in Uganda have the potential to reduce greenhouse gas emission by 64% compared to Uganda's conventional farms. It is generally believed that more research is needed on the potentials and challenges of organic livestock production.

CANADA: The organic farming movement in Canada emerged in 1950s and significant development occurred in 1970s. In this period, organic farmers formed organizations in six provinces. In 1974, McGill University founded an Ecological Agricultural Program that serves as an information clearing house for organic farming. Certification bodies were formed in the 1980s. In 2003,

Canada had 3,100 organic growers, or 1.3% of its agricultural producers farming on 390,000 hectares. The industry has grown at about 20% per year. Organic products account for about 90% of its dairy, 20% of its fruit and vegetables and 10% of the nation's grain crops. Saskatchewan has the greatest number of organic growers at slightly less than 4,000 or about 1.8% of Canadian farms.

The Canadian organic farming standard was approved in 1999 by the Standards Council of Canada. It was developed jointly by the Canadian General Standards Board and COAB. The standards are voluntary and do not constitute a minimum specification for organic production. This creates three issues: 1) a lack of uniformity of organic products, 2) the loss of access to certain export markets and 3) confusion among consumers. Grower perception is that organic systems are 30-50% less productive, more labor intensive, receive a premium price and emit less carbon dioxide compared to conventional systems.

Between 2006 and 2011 there was a decline in the number of Canadian farms. Interest in organic farming, however, increased. It was hypothesized that this was due to increasing consciousness of health, quality of farm products and environmental quality. The availability of organic products remains low and distribution systems are undeveloped. Consumer demand, however, is strong, particularly in urban areas. To facilitate better organic and ecosystem-based farming, additional training programs are needed. Organic and ecosystem-based agriculture research and development projects exist at the national and provincial levels. Professor Bird's experience indicates that Ontario has an exceptionally strong soil health research initiative that involves collaboration among universities, the Provincial Ministry of Food and Agriculture and Agriculture Canada. Farmer participation in educational workshops is high.

ITALY: Among the European Community countries, the status of organic farming in Italy ranks second (1.2 million hectares) after Spain (1.6 million hectares). The national demand for organic products grew by 7.5% in 2013. The total land area in Italy is more than 29 million hectares of which 37% or about 11 million hectares were under annual or permanent crops. The majority of the farms were less than three hectares in size. Reports on Italy's organic agriculture indicate that it continues to be a major player in the European organic market for fruit and vegetables (Campagnoni *et al.* 2000). Italy has about 50,000 organic producers. As reported by the Ministry of Agriculture, the number of organic farms in Italy increased by 3% and the amount of land under organic farming system increased by more than 6% in 2013. This has been possible because of the prevailing favorable climate, soil, eco-systems, technological supports, appropriate package of practices and close geographic access to major export markets. In addition, there is high demand of organic quality foods among domestic consumers. In general, people are concerned about overall quality of the environment. Consumers, particularly in north, prefer to buy fresh organic fruit, vegetables, honey, jam, eggs, olive oil and yogurt at premium prices. They shop at supermarkets and traditional shops or purchase directly from farmers. There are about 1,270 shops in Italy that specialize in organic food products. More than 50% of fruit farms are located in Sicily. Vegetable production follows the same geographical pattern in the southern regions of Sicily and Apulia.

The domestic demand of organic food in Italy is high and cannot be met by local producers. An import market has evolved with organic produce coming from Egypt, Cameron, Angola, South Africa, Turkey, Israel, Argentina, Colombia and Peru. This includes onions, carrots, potatoes, peppers, cucumbers, kiwis, apples and pears. It is estimated that about 17,000 tons of organic products are imported each year. A large share of domestic fruit and vegetables are exported to European Community countries, mainly northern Europe and Switzerland. Yields of sugar beets, tomatoes and vegetables have increased significantly, but wheat yield has declined. Since, 1998, the organic sector in Italy has developed significantly. It has the largest area farmed organically and the highest number of organic farms in Europe. Fruit orchards, especially citrus, receive high subsidies while other horticulture crops are not policy supported like cereals and fodders.

The Italian Association of organic farming (Associazione Italian Agricoltura Biologica-AIAB) was established in Turin in 1988. It was designed to organize farmers, researchers and consumers to assure progress in organic farming in Italy. Eighteen regional chapters gradually evolved. They are coordinated by a federal office located in Rome. It supports development of services to producers and also collaborates with government agencies and research institutions in standardization of the methods. The AIAB also specializes in evaluation and certification of organic brands and labels, as well as arranging vocational education and training for the interested growers. Organic farmers in Italy are also certified by other bodies, such as, BIOS, CODEX, CCPB, BIOAGRICOOP and IFOAM Italy. In June 2007, the European Council of Ministers agreed on a new Council Regulation on organic production and labelling. In this process, greater emphasis is placed on environmental protection, biodiversity and high standards of animal protection. This resolution is designed to ensure consumer confidence and interests. Accordingly, food may only be labeled as organic if a minimum of 95% of agricultural ingredients and inputs are organic. Although Italy has faced financial challenges, it continues to be a leader in the domain of organic trade and exports.

MEXICO: Mexico has 198 million hectares of land. Fifteen percent of the land is zoned for agricultural crops and 58% for livestock. About 20% of the crop land is irrigated and organic farming is practiced on 0.4% of the land. Historically, agriculture has been an important sector in the nation's economy and politics. During the Mesoamerican period, the domestication of plants such as, maize, beans, pepper, tomato, squash, avocado and cacao took place in this region. Turkeys were the main domesticated animal prior to the Spanish introduction of cattle, horses, donkeys, goats and sheep. Mexico currently produces corn, wheat, tropical fruit and vegetables. About 60% of the agricultural products are exported to the U.S. During the second Green Revolution, Norman Borlaug, with support from the Rockefeller Foundation, developed dwarf varieties of wheat and maize with disease resistance and high protein content. This significantly increased yield.

At the beginning of the 21st century, the most profitable organic products were coffee, sugarcane and cotton. Mexico also produces organic beef, fruit, vegetables, corn, milk and poultry. In 2007, 3.2 million tons of vegetables and 1.8 million tons of fruit were exported to U.S., including bananas and papayas. In 2005, the Rodale Institute, reported that although the demand of domestic organic products is small, the land area under organics is expanding at 45% per year, mostly by small farmers having a farm size of just over two hectares. The incentive is the premium price for export produce. Ninety-eight percent of the country's organic products are exported.

Organic farmer markets have been started in major cities (Gonzalez and Nigh, 2005). Organic production in 2004 was estimated to be $350 million on 300,000 hectares, which was three times that of four years earlier. Organic coffee accounts for 10% of all coffee land in Mexico. This is greater than the value of all other certified organic crops including vanilla, mangoes, African palms, pineapples, papaya, oranges, field beans, avocados, soybean, and cacao. Germany, Holland, England, Japan and Switzerland are key buyers.

About 50,000 small farmers, mostly coffee growers, of Oaxaca and Chiapa, produce over two thirds of Mexico's organic products. Certification is done by farmers in groups or their cooperatives, since individual farmers may not be able to meet the requirements of organic certification due to the small size of their farms. There are about 18 organic certification agencies. OCIA (Organic Crop Improvement Association), however, is the primary certifier. For export, the producers or groups have to follow the certification standards of the respective import nations, such as the U.S.A. or E.U. Generally, inspection is done by an outside inspector on an annual basis. Grower incentives are strong. If one farmer of a group or their cooperative is found not to be following the standards, the whole group can lose its certification for a period of one to three years.

U.S.A.: Following communications between J. I. Rodale and Sir Albert Howard, the first issue of Organic Gardening magazine was published in 1942. The Rodale Institute was formed by Robert Rodale to facilitate development of viable organic farming in the U.S.A. Professor Bird had the honor of serving on the Rodale Board of Directors for nineteen years. Although the National Organic Program was authorized in the 1990 Farm Bill, the National Organic Standards were not implemented until eleven years later. This was facilitated under the leadership of Dr. Kathleen Merrigan, with major input from the entire U.S. A. organic community.

In 2020, organic agriculture in the U.S.A. is more than a $50 billion industry. There are more than 15,000 certified farms growing organic produce and livestock on more than five million acres. This, however, is less than 1% of the more than 900 million acres of land used for agriculture. Most of the organic produce is consumed domestically. The organic farm industry is divided into two groups, small-scale farms that market locally and large industrial-scale farms that market through chain stores. Most states have organic farming associations. Some states have laws governing organic certification agencies, while most use organic certification organizations approved by the National Organic Program.

While interest in food quality and environmental health has increased on a national basis, organic food is significantly more expensive than that produced under conventional agricultural practices. Professor Bird became well aware of this when he did his family's weekly shopping following the quarantine restrictions associated with COVID-19. In 2020, many conventional farmers are aware of the issues associated with soil health degradation. They have become very interested in learning about the organic farming practices that have the potential to assist them in soil health maintenance and remediation.

Since organic agriculture is a dynamic system, one of the best ways to keep abreast of current developments in the U.S.A. is through on-line websites. The National Organic Program can be accessed at https://ams.usda.gov, the Rodale Institute at https://rodaleinstitute.org and Real Organic at https://www.realorganicproject.org. To have a comprehensive understanding of the global nature of organic agriculture, ecosystem-based agriculture, global food security and zero-hunger, it is essential to have an understanding of the role of rice.

Rice

Eco-Geography of Rice

Ecosystem-based rice production is an eco-friendly, cost effective, sustainable, health promoting, nature loving and organic resource conserving system. Rice (*Oryza sativa*, L) is native to the deltas of great rivers, Ganges, Chang (Yangtze), Tigris and Euphrates. Since rice was incorporated into human diets 5,000 years ago in the Indus Valley of Indian sub-continent, its production technology has changed significantly. Rice occupies a highly unique ecological dimension that is unlike other grassland agricultural systems. It is cultivated in medium to low land water-logged conditions, mostly in high rain fall areas. About 88% of rice grows under flooded conditions during part of, or the entire cropping season. The total land occupied by rice is 10% of the world's arable land. Most rice is cultivated under conventional agricultural methods, using synthetic fertilizers and pesticides for high yields and short-term food security. Only limited amounts are grown by small or marginal farmers. Some of these farmers are not socio-economically solvent and a few live in remote tribal-areas. These farmers grow rice using traditional practices and local organic resources. An exception is the cultivation of fine, long slender, scented rice varieties used for domestic and export premium price/profit markets.

Rice is grown under different environmental conditions including upland, dry land and flooded or puddled rice soils. After flooding or puddling rice soil, its physical, chemical and biological attributes change. The changes include structure, redox potential, pH, nutrient availability, photo-chemical behaviour and micro-biological characteristics and activities. Traditional wetland rice culture has been sustainable for thousands of years. This is due, in part, to soil nutrition related to biomass associated with natural algal growth allowing bacterial nitrogen fixation that results in moderate yields. Rice, apart from being a nutritious primary staple food for more than 50% of the world's population, is used as a constituent of many products. These include animal feeds, baby foods, breakfast cereal, rice noodles, rice snacks, takumi, chicken rice, fiesta rice, rice bran oil and rice straw boards. The production of high-quality organic rice is an important issue in overall rice culture. To combat iron, and zinc malnutrition among women and children in Afro-Asian countries, biofortified non-GMO rice varieties have been developed to assist in resolving these issues. In addition, the food nutrition value of brown rice is known to be superior to that of white rice (Table 4.11).

Rice is grown on 161 million hectares or 18% of the land base used for agriculture. India has the most with 45 million hectares and China second with 29 million hectares. North America produces seven million hectares and African nations nine million hectares. The major areas under rice are rain fed (2,500 to 11,000 mm/year). Others are partly irrigated, or occasionally flooded/inundated. High rates of soil erosion and loss of nutrients through run-off, seepage and leaching are often associated with rice production. This results in the need for a special type of holistic management for optimal rice productivity. Areas under dry land or upland rice cultivation in plains or hill ecosystems are limited A new vision and technologies for rice production are required for maintenance or renovation of soil health and global food security. Many of the so called proven technologies of biological-based agriculture, like growing vegetable crops, fruit crops, mixed cropping of rice-legumes, animal husbandry, bee keeping are not practical for rice production. Development of alternative systems are especially important for the highly populated rice belt of the world where growers are mostly small and marginal. Yield stability is of primary importance for these small and fragmented farms that frequently have limited natural resources.

Table 4.11. Nutritional composition of white and brown rice (www.healthline.com).

Component	White Rice	Brown Rice
Protein	1.42 g	1.83 g
Total lipid	0.15 g	0.65 g
Carbohydrate	14.84 g	17.05 g
Fiber	0.2 g	1.10 g
Sugar	0.03 g	0.16 g
Calcium	5 mg	2 mg
Iron	0.63 mg	0.37 mg
Sodium	1 mg	3 mg
Sat. fatty acids	0.04 mg	0.17 mg
Total fatty acids	0 g	0 g
Cholesterol	0 g	0 g
Calorific energy	68%	82%

As a holistic, biologically active, sustainable management system, ecosystem-based rice production is designed for the welfare of humankind and their biotic and abiotic resources. This includes the principal crop of rice, but other crops grown in sequence. The associated treatments or materials applied to the soil or crop as nutrients or pesticides need to have long-term synergistic positive impacts on both the soil and crop (Pornpratansombat *et al.*, 2011).

When paddy fields are flooded by excess rain, adjacent organic farms can be chemically contaminated through run off or seepage water. Planting barrier crops such as forage grasses, shrubs, Sesbania or *Eucleana mexicana* can be used to protect organic rice production sites. In addition, organic farming policy and methodology must be integrated with crop rotations such as rice-jute, jute-rice or wheat, potato, oilseed crops or pulses for the system to have positive human, animal and environmental health impacts. The systems must be designed to be ecologically-friendly in regards to the soil, water and atmospheric resources that support beneficial insects and soil-borne microbes.

Rice is the staple food for more than 50% of the global population. Its annual production is a key to global food security. Rice is particularly important for densely populated Asia that accounts for 90% of world's rice production and consumption as the main dietary food. India has about 17% of world's human population and 11% of livestock on a limited land area of 197 million hectares of agricultural land. In addition, it only has 2% of the water resource and conservation practices are poor. In 2014, food production in India was about 264 million tons, harvested from 43 million hectares of land. India needs to produce an additional two million tons of rice annually to assure food security. While this is not possible immediately, ecosystem-based organic farming is a key to India's food security. In addition, with India and China being the most populated countries in the world, their food security will require a 4% to 5% annual growth in agriculture to attain the additional 8 to 10 million tons of food grain required annually to keep pace with population growth. For food security, India will need to produce about 280 million tons of food grains, 50 million tons of oil seeds, 20 million tons of pulses, 130 million tons of vegetables and 85 million tons of fruit in the very near future. Therefore, whatever technology is used for rice production, it will be a significant challenge under diminishing natural resources and a changing climate. Food security is only achieved when all people at all times have physical, social and economic access to adequate, safe and nutritious food for their dietary needs and life style. Most Afro-Asian nations are far behind the goal of food security. This is especially true for rice production.

Most rice production in the Indian subcontinent, China, Bangladesh, Vietnam, Thailand, Philippines, Cambodia, and Indonesia employs conventional agricultural methods (Table 4.2). This results in degradation of the quality of soil, water and air resources, resulting in nitrate accumulation in pond and river water, in addition to arsenic, selenium and fluoride toxicity. All of these have negative impacts on local ecosystems and result in significant environmental and health problems.

Table 4.12. National rice consumption in millions of tons in 2018.

China	14	India	97
Indonesia	37	Bangladesh	35
Vietnam	27	Philippines	12
Myanmar	11	Thailand	10
Japan	9		

Children and women are affected the most. There is a need for urgent attention, amelioration and mitigation, while paying appropriate attention to food security.

In addition to India and China, rice is also an important crop in Africa, Latin America and Caribbean nations. In the U.S.A., rice is grown in Arkansas, California, Louisiana, Missouri, Mississippi, and Texas. The crop consists of 73% long grain, 26% medium grain and 1% short grain rice (www.ipni.net). In Arkansas and Mississippi, a soybean-rice rotation is common, although monoculture is also practiced. Conventional tillage, fertilizers at 135-235 kg N/ha, dry seeded, delayed flooded to the 5 to 6 leaf stage or water seeded with pregerminated rice systems are practiced. The introduction of Provisia, a non-transgenic herbicide resistant variety of rice, by BASF has gained acceptance. The pests and infectious diseases are managed with resistant rice varieties and IPM methods.

Brazil, Columbia, Equator, Peru, Argentina, Uruguay, Guyana and Paraguay grow both irrigated and rain-fed rice. Brazil produces more rice than the other seven nations.

Most rice grown in South America is exported, with some being used for domestic consumption. The trend of increased rice consumption in China, India, U.S.A. and South America supports the view that rice is the major staple crop for global food security and rural economic prosperity.

In 2006, African nations produced about 9.0 million hectares of rice on an annual basis. Most of this was *O. glaberrima* and *O. glaberrima* x *O. sativa* varieties. Yields are low, but expected to increase at 7% per year. About 75% of the rice production is rain-fed and 25% irrigated. The gap between rice production and consumption is high. Most of the rice consumed is imported at a cost greater than $1.4 billion per year.

All types of agriculture require ecosystem disturbance. Physical, chemical and biological disturbances can result in environments favorable for pests, infectious diseases and soil health degradation. It is important, therefore, to keep all forms of physical, chemical and biological disturbance at the minimum level that will result in the desired crop quality and productivity. When this is achieved in ecosystem-based agriculture, the system will be self-regulating and sustainable. When a rice system is out of balance, there will most likely be a need to implement pest management strategies and tactics. While avoidance should always be the first consideration, containment and control are often the only options. Proper use of biological, chemical, physical, cultural, genetic and regulatory pest and disease tactics requires understanding of the ecosystem of concern. Without this knowledge, implementation of these tactics can result in a bifurcation (turning) point, leading to long-term low rice yields and poor rice quality.

Organic Rice

Various types of ecosystem-based rice farming with organic and inorganic resources are practiced by small and marginal farmers throughout Asia, Africa and Latin America. This should be considered as a pre-transitional phase to certified organic rice production. The escalating costs of conventional farming, low profit margins, low quality products, human health issues and environmental concerns should serve as a catalyst for a global transition to ecosystem-based and certified organic rice farming.

As previously indicated, medium and coarse grain rice are produced for domestic consumption and export under conventional and traditional farming practices. A significant portion of this is grown on small farms with limited resources. These characteristics need to change. Organic rice production has the potential to have a significant positive impact on this badly needed change.

Rice has a wide soil and climate adaptability. More than 5,000 varieties are grown. Rice is produced from hill regions to the sea coast and can be grown in soils with pH ranging from 4.8 to 8.5. Rice cultivation can be followed by crops such as groundnut, jute, green gram, rape, wheat, black gram or okra that provide the biological diversity for high soil organic matter and biological nitrogen fixation needed for organic rice production. Very little long-term research, however, has been done on organic rice production under different soil, rainfall, socio-economic and cropping systems. This is especially true for weed, insect and disease management. As exception is fine scented rice. Production of this type of rice, however, is negligible.

The demand in Europe and North American for Indian organic long, slender aromatic Basmati rice is increasing. Some parts of India are ecologically suitable for growing Basmati rice. The aroma of Basmati is due to the presence of more than 100 volatile compounds. The principal compound is 2-acetyl-1-Pyrroline (2AP). In addition, some Indian short grain, fine scented varieties such as Aijon, Joha, Bhutmuri and Govindabhog, are grown organically for both domestic and foreign markets. (Nadaf and Wakte, 2006). The nutritive value of Basmati rice is superior to short grain white rice varieties, with a calorific value of 349, 8.1% protein, 77.1% carbohydrates, 0.6% fat and 2.2% fibre.

High quality scented Basmati rice grows very well in India in rich soils with high organic matter. Parts of Haryana, Punjab, Himachal Pradesh, Uttarakhand and Uttar Pradesh are excellent areas for Basmati rice production. In 2010, organic Basmati rice was grown on 1.1 million hectares in India (www.biocrop). Although production is less than coarse grain rice grown by conventional practices, farmers receive a premium price resulting in more profit from organic scented rice, compared to rice from conventional production systems.

Cultivation technologies and other management practices for organic rice differ. They must, however, be based on local soil, water, biological resources, and climatic conditions. Soil health, water resources, weed control, availability of sufficient organic resources or animal manures and favourable socio-economic conditions are essential for successful organic rice production.

The amount of upland rice area is limited on global basis. Organic rice, however, can be successful under high and hilly dry land conditions with direct seeding, utilization of local organic resources, crop rotation and mixed cropping rice with legumes under mostly high rainfall areas. This system is often used by tribal people. Land levelling, in-situ decomposition of weeds, green legumes, and rice straw with rotary hoeing is part of the process.

Farm-yard manure, compost, rock phosphated compost, vermicompost, oil cakes, bio-fertilizers, BGA, and Azolla, are used as sources of nutrients in organic upland rice systems. Vermicompost or compost with saw dust plus *Tricoderma* and *Pleuratus* stimulate organic matter decomposition. In 1990, Professor Chakraborty found that the low cost technology of using farm weeds, rice straw,

119

water hyacinth, and other crop residues composted with cow dung, rock phosphate/basic slag (60-100 mesh size) and treated with *Phosphaobacterin* can be useful in rice systems.

Aspergillums, Azotobacter, Trichurus spiralis or their *in situ* decomposition prior to planting, improved soil fertility and crop productivity. In addition, it is known that when green refuse is applied to rice soil in conjunction with basic slag/rock phosphate in different cropping sequences, it improves soil fertility more than inorganic fertilizers. Baishya *et al.* (2017) demonstrated that a Toria (*Brassica rapa.*), wheat, rice, okra system resulted in the greatest rice yield and enhanced soil enzymatic activity. This was true under irrigation for dehydrogenase, phosphatase, and urease when the nutrition was supplied from organic resources such as farmyard manure, vermicompost, oil cake, Azospirillum and Azotobacter. In the acid soils of Assam, rain fed rice followed by a Toria-black gum (*Vigna mungo*) system resulted in a high yield of nine tons/hectare. Organic manuring with 33% farm yard manure compost, 33% vermicompost and 33% mustard seed cake, with a biological pesticide made from Pumello fruit, cut into pieces and applied to soil at 30 kg/ha, was successful.

Under average field, soil and water conditions of limited resource small farms, it is difficult to achieve sustainable food security growing only organic rice. A biologically diverse system is essential for transitioning to sustainable productivity and biologically healthy soil. The use of bio-fertilizers with Azolla, BGA, Azospirillum, Azorobacter, Phosphobacterin and rock phosphate fix atmospheric nitrogen and add a considerable quantity of organic matter and available phosphate to the soil. Roger (1995) discussed the prospects and problems of nutrition of wet land rice production using heterotrophic bacteria. Use of Azolla, and green manuring with legumes such as *Sesbania aculeate* or *S. rostrate* are considered as the best way to increase the fertility of organic rice soil and yield. Pulses or fodder can be used as a substitute for legumes in wet land rice soil production. Including legumes, however, is recommended for strip cropping or for the boundaries of wet lands in upland organic rice. Green manuring is a good practice for organic rice cultivation. This is well-known in traditional agriculture, but due to socio-economic reasons, farmers prefer to plant a cash crop before rice. Azolla culture, and green manuring is recommended for deep water rice or alley cropping. Legumes such as *Glyricidia, Euchleana* and *Tephrosia,* in surrounding rice fields have high potential for facilitating appropriate nitrogen and organic soil matter.

Plant protection from insect pests and disease pathogens in organic rice cultivation requires biological management practices. In general, this is more difficult than using synthetic pesticides. Pest populations must be maintained beneath their established damage thresholds. This has to be done in an appropriate cost-benefit manner using the fundaments of integrated pest management. Tillage and use of resistant varieties are options. Puddling or sunning soil, cultural, direct population reduction with pesticides approved for organic production, sea-weed, leaf extracts, bio-pesticides and bio-stimulants are available. According to FAO in 2006, 16 million tons of sea weeds are used annually for plant nutrition and as bio-stimulants to induce resistance against insects and diseases. Use of resistant varieties, crop rotation, strip cropping, barrier cropping, and light traps are important options for insect and disease management in organic systems. They are, however, knowledge intensive practices. Flooding and puddling of soil and integrated organic rice farming with duck and fish culture is an attractive option for control of insects, soil borne pests such as nematodes and other pathogens. A strong organic vibrant soil base with favourable ecosystem components is known to have less risk to damage from insects and diseases than systems using synthetic fertilizers.

In low land flooded areas of Bangladesh, Thailand and India (Tamilnadu, Kerala, Assam), China, Laos and Vietnam, cultivation of deep-water rice integrated with duck and fish culture is an excellent system for economically successful organic agriculture. Integrated rice-duck or rice-duck-Azolla-Fish farming is practiced in U.S.A., Japan and Philippines. The integrated-rice-duck-fish (IRDF) system was developed by Takao Furuno in 1989 in Fukuoka Prefecture, Japan. This method is a biological system that eliminates the need of fertilizers and herbicides. The ducks effectively control weeds and harmful insects. Their paddling movements appear to stimulate plant growth and duck manure enhances the soil nutrition (www.equtorinitiative.org). Professor Bird visited the Furuno farm in the mid-1990s. It was a truly amazing system. The fish in the rice paddy were abundant enough for Professor Bird to easily capture them with his hands. In addition, the sky was full of birds that had not existed when the farm produced rice under conventional agricultural practices.

Historically, a modified version of IRDF was practiced by small farmers in East Asia. Ducks are used in small ponds adjacent to the rice fields managed under traditional agricultural practices. This is not, however, practiced widely in India and other nations on a commercial basis. Utilization of aquatic weeds such as water hyacinth and their decomposition *in situ* in rice fields provides biological nutrition in a rice rape seed system for nutrient management practice in hilly areas of Meghalaya. Application of farmyard manure at 2.5 tons/hectare plus *Eupatorium odoratum* at 2.5 tons/hectare or *Alnus nephalensis* at 2.5 tons/hectare is effective and results in yields equivalent to that of inorganic fertilizers. Weeds contain high NPK (2.3 -2.4% N, 0.8-0.9% P and 1.2% K).

The northern Himalayan and north-east hilly regions of India are ideal for upland organic rice and other crops grown in an organic crop rotation system. These regions have a great diversity of flora and fauna with good biomass resource and high rainfall (2,000-11,000 mm/year). Crops like potato, beans, carrot and okra grow well in raised terraces and rice in sunken beds utilizing the available green biomass for soil nutrition. Optimal organic rice yield was obtained following early manuring. Reports from South Korea showed that the use of compost at 12 tons/hectare, plus rice straw at five tons/hectare, hairy vetch, rock phosphate, powder of lime magnesia and ash resulted in good yields of rice. While this system maintained soil fertility, crop productivity was a little lower, compared to the conventional management system.

Integrated plant nutrient systems for rice have been reported by scientists in various countries. It is an effective way to have good productivity, while maintaining the biological and physical properties of soil (Patra *et al.*, 2017). Organic sources of nutrients supplied to the crop preceding rice, benefit the succeeding rice or other crops. Application of 50% of the recommended rate of fertilizer (NPK at 100-22-41.3 kg/hectare) coupled with 50% of the recommended rate of nitrogen through green manuring with *Sesbania cannabina* in the rainy season, followed by 100% RDF applied to summer rice can result in highly respectable rice yields over a long period of time.

A system of rice intensification was developed by the French Jesuit Father, Henri de Laulanie in Madagascar in 1983. It has been adopted worldwide under the leadership of Dr. Norman Uphoff of the Cornell University International Institute of Food and Agriculture. By 2013, the system of rice intensification was used by about 10 million rice farmers in 54 countries in Asia, including China, India, Sri Lanka, Philippines, Malaysia, and Vietnam. The system has resulted in record rice yields in China and 50% to 100% yield increases in India. The primary features of exceptional yields of rice are: 1) use of high yielding long-duration hybrid seeds, 2) planting 8-12 day-old seedlings in wide spacings at 1 to 3 seedlings/ hill, in 30 x-30 or 50 x-50 cm, 3) scheduled irrigation with 5 cm depth, after 3 days after application, 4) frequent weeding, at least twice, preferably by cone/rotary hoe and 5) careful levelling and intensive land preparation, This method is used by organic rice growers in several nations.

Weed control under organic dry land rice cultivation during the early period of growth is critical. This is also true for upland and dryland direct seeded rice. Mechanical methods, preferably rotary hoeing is recommended to incorporate the weeds into soil at a low cost, while supplying organic matter and nitrogen. A mixed vegetation of weeds such as *Echinochloa colonum*, *Digitaria* spp., *Cyperus exaltatus*, *Commelina nudiflora*, *Ludwigia parviflora* and *Vandallia crustacean* in early vegetative stages of growth can remove 52 kg of nitrogen/hectare. In addition, the nitrogen content of weeds can be 4% higher than that that of adjacent rice plants.

Transitioning to certified organic rice culture from conventional agriculture, is not immediately possible due to many constraints linked with climate, ecology, availability of organic resources, market linkage, local dietary requirements and the socio-economic status of many farmers. The demand for organic food, including rice, is increasing significantly in Europe and North America. Domestic markets are also increasing. This, however, is based on the demands of an elite class of people in urban areas. Although relatively little research has been done on organic rice production, it is the ultimate stage of farming in harmonious equilibrium with nature. Ecosystem-based rice production is an appropriate intermediate transition from the conventional system. It should provide positive attributes for future rice production on a global basis.

While production of organic rice on a global basis has been slow to develop, organic farmers have found success in growing fine, slender grain scented rice varieties for export markets at premium prices under contract systems. India is the leading country for growing and exporting Basmati rice grown under organic farming practices. It is apparent that government support systems would significantly enhance organic rice production.

Since the rate of growth of conventional rice production is decreasing, ecosystem-based and certified organic rice production are essential for meeting the needs of global food security. The target areas for rice should be the humid tropics where rain fall is high and bio-diversity is adequate. Biomass resource availability is essential for small and marginal farmers since commercial fertilizers and pesticides use is limited or negligible. Yield stability, however, is a significant challenge because of the many physical, ecological, ethical and socio-economic limitations associated with global rice production. The future of organic rice production is strongly integrated with the future of food security and hunger.

Chapter 5. Food Security and Hunger

A life cannot bloom and proceed without food. A hungry person cannot be productive or listen to advise from an elder with an empty stomach. Food is a fundamental right! A strategy for achieving zero human hunger is a global imperative. According to the U.N. Charter on Food Policy, *there is one nest, our biosphere.* Following independence in 1947, the first Prime Minister of India, Pandit Jahwarlal Nehru, emphatically and rightly remarked, one nest in Upanishad is *Yatra Viswyam bhabetya Eko-Neerham (....the world is a single nest, common shelter for all, living together in harmony for peace and general welfare, sharing each other for all times to come.*). In the 21st Century with more than 7.8 million people living with limited high-quality natural resources, this statement is still true today.

The question is, which of the five types of agriculture; ancient, traditional, conventional, eco-system-based or organic are best suited for a successful strategy for achieving zero-hunger by 2030 and maintaining it for the long-term? The answer is that no one of these systems can achieve the zero-hunger goal on a short-term basis. Achieving zero-hunger is a long-term process. Based on the history of change, the process will take more than a decade. It will require a planned and most likely a regulated transition period. The best option for achieving zero-hunger is through ecosystem-based agriculture, with organic agriculture as a specialized type of environmentally-sound food, feed and fiber production.

Global transition to ecosystem-based agriculture on a short-term or long-term basis is complex. It mandates significant investments in education, facilitation and persuasion. The fact that farming is a complex, open system with self-organization, emergent properties not present in its parts, nestedness and bi-furcation points, make the challenge even more difficult (Wessels, 2013).

Extensive debate on global food security and zero-hunger has taken place at national and international levels. With the current dominant hypothesis being that global human population will increase to 9.7 billion by 2050, Nehru's statement that *everything can wait, but not agriculture*, becomes even more critical. As a scientist that believes that alternate hypothesis testing is the basis of scientific method, Professor Bird frequently asks, his students, *what is the alternate hypothesis to the 9.7 billion people hypothesis? Could it be 3.0 billion, the global population when I was an undergraduate college student?* Such a tragedy is totally incompatible with the Upanishad philosophy and all others of the civilized world. COVID-19, however, is a reminder that the Bubonic Plague and other unknown pandemics likely kept human population density low for thousands of years. The solution to the challenge of zero-hunger must be based on: 1) meaningful dialogue on a global basis, 2) the fundamentals of science and 3) implementation of appropriate technologies.

Both ecosystem-based and organic agriculture will play key roles in the future of global food security. About 95% of agricultural land in Europe and North America is farmed using conventional technologies. In 2020, a significant number of conventional farmers are interested in the technologies of ecosystem-base agriculture, while maintaining their productivity and profit levels. In addition, the demand for organic foods is increasing rapidly in Europe and North America.

What is zero-hunger? Hunger is a feeling caused by a lack of food and a desire to eat. It is a physiologically exhausted condition of the body resulting in a craving for food. Hunger can be a short-term or long-term phenomenon. A long-term lack of accessibility to food or inability of the body to properly process food results in malnutrition. This involves an improper amount and balance of proteins, carbohydrates, fats, calories, vitamins and minerals. A lack of adequate fresh drinking water is also a component of the malnutrition equation. Malnutrition may be under-nutrition or over-nutrition. Under-nutrition is most frequently related to poverty or infectious diseases.

The World Food Summit of 1996 defined food security as the condition when *all people, at all times, have physical and economic access to sufficient, safe and nutritious food that meets their dietary needs and food preferences for an active and healthy life.* Zero-hunger and food security, therefore, are synonymous. United Nations Secretary General Bau-Ki-Moon launched a zero-hunger challenge in 2012 at the Sustainable Development of Agriculture Conference (Rio-2) in Brazil. All of the active nations agreed that hunger, food insecurity and malnutrition should be ended by 2025. In general, this has resulted in a global challenge of 100% access to food for all.

Is zero-hunger possible? Although enough food is produced on a global basis for zero-hunger and new agricultural technologies and scientific discoveries are continually becoming available, the goal of zero-hunger has not been achieved. In 2006, IFOAM reported that the major constraints to global food security are distribution and socio-economic/political policy. Both of these need to be addressed immediately. Since more food is produced in the world than needed for food security, simply increasing global food supply will not solve the problem. Amartya Sen, the 1988 Nobel Memorial Prize in Economic Sciences award winner indicated that while *starvation is the characteristic of some people not having enough food to eat; it is not the characteristic of there not being enough food to eat.*

What are the realities of global hunger? The United Nations Agriculture Organization estimated that in 2014-2016, 795 million people suffered from chronic undernourishment. Most of these individuals live in the developing countries of Sub-Saharan Africa and, East and South-East Asia. This represents about 13% of the population of these nations. According to FAO in 2015, only about 11 million people are undernourished in the high-income countries of North America, Europe, Mexico and Australia. Asia is the center of hunger and poverty, with two out of every three people being undernourished. Paradoxically, there are more than one billion people in the developed and industrialised world with health problems resulting from over consumption of food. FAO estimated that if just one fourth of the food lost or wasted were utilized, it would be enough to have a significant positive impact on alleviating food security issues.

The losses and wastes of processed and non-processed food is a significant factor in food insecurity. In India, an estimated 60,000 core rupees are lost annually due to wastage of food in hotels, restaurants and reception parties. In addition, about 21 million tons of wheat grain are lost every year due to a lack of storage infrastructure. About 15% to 20% of the fruit and vegetables are lost due to a lack of processing and refrigeration facilities. Thus, food security or elimination of hunger in India cannot be achieved solely through India's subsidized food program. Reducing food wastage is an essential part of the food security solution. Annually, India must produce an additional 62 million tons of food grains to meet the National Food Security requirements.

It is inappropriate for individuals and nations to waste food and engage in overconsumption while approximately 10% of the global population is hungry or undernourished. In Germany, diners are fined one or two Euros if there is any intentional or incidental food wastage on their plates. In Switzerland, customers are charge an additional five francs for food wastage. India's 2013 Food Security Bill has no provision of penalties for food wastage. Penalties for wasting food is symbolic and designed to draw public attention to food security, hunger and malnutrition. In Switzerland, however, about two million tons of high-quality food is wasted each year. In Denmark, Sweden and U.S.A., food products not eaten before their expiration date for safe consumption are often discarded as waste. In the U.S.A., the amount of discarded waste food is equal to about 10% of what is consumed on an annual basis. All of these food system residuals have significant potential for use in various ecosystems. In true ecosystem-based system there are always residuals. They should never be considered as wastes.

In Sub-Saharan Africa, there is an acute concern about the increasing rate of hunger and the number of malnourished children (Von-Braun, 2005). While this exists, high social classes in affluent regions consistently have surplus production and surplus consumption. It has been estimated that as much as 20% of the global population live in a state of constant undernourishment. This is an indication of very slow progress towards zero-hunger.

According to the World Hunger, Poverty Facts Statistics of 2016, the following eight nations are among those suffering most from food insecurity and malnutrition.

- Chad (13 million people, 2.4 million with food insecurity).
- Zambia (48% undernourished).
- Timor-Leste (1 million people, with more than 50% of the children are under-nourished).
- Sierra Leone (15% of children under five years of age suffer from malnutrition.
- Madagascar (36% percent of the rural population do not have food security).
- Afghanistan (33% of the population is food insecure).
- Niger (60% live in poverty and 44% of the child suffer from malnutrition).
- Yemen (Female headed-households are more food insecure than male-headed households and children malnutrition is highest in the world.

FAO estimated that if the challenge of global zero hunger is achieved, the lives of about 2.8 million children under five years of age could be saved each year through an adequate supply of nutritious food. Food hunger in Southern Asia (India, Bangladesh, Pakistan) is about 16% of the population, mostly because of poverty among rural people. In eastern Asia, mainly China and South East Asia, nations such as Indonesia, Philippines, Myanmar and Vietnam should have the capability to substantially alleviate the malnutrition by about 10%. The same should be true for Latin American nations. Despite high economic growth attained recently in India, malnutrition accounts for about 54% of child deaths. Harvard University Professor Amartya Sen remarked that what is sometimes called protein-energy malnutrition is nearly as high in India as in some Sub-Saharan Africa nations. In addition, iron deficiency and anemia are high among pregnant women in India (Swaminathan, 2019).

Food security is a multidimensional concept. It depends on the availability of food through increases in local and national agricultural productivity and imports. A second factor is the accessibility of food based on people's buying capacity and if they can procure it in a timely manner that fulfills their dietary habits. A third dimension is whether the rural or urban population can consume and digest the food in a manner related to the available fresh drinking water. Inadequate fresh drinking water is often a key limiting factor in resolution of the hunger and malnutrition challenge. The Food and Agricultural Organization states that food security occurs when *all people at all times have sufficient physical, and economic access to safe and nutritious food to meet their dietary needs and food preferences for an active and healthy life.* A Global Food Security Index (GFSI) was formed in 2012 from 35 factors. Use of the index clearly indicates that high income nations like the U.S.A., Denmark, Norway, Switzerland and Germany have high GFSIs. (Table 5.1). Developing nations like India, Pakistan, Nepal and Bangladesh have significant incidences of hunger and malnutrition, especially in respect to the poor classes of people in urban and rural areas. It is estimated that about 2.1 million children below the age of five die annually from food deficiency issues.

Three of the indicators used in the GFSI are availability, affordability and quality/safety (Table 5.2). The 2019 GFSI score for India was 58.9 compared to 45.0 in 2012. Ittyerah

(2013) reported that food security in India is of prime importance because more than 33% of the population of India are poor and about half of the children suffer from malnutrition, especially vitamin A deficiency. A 2008 United Nations Development Program survey of 16 districts of India found that about 7% of households have inadequate access to food. A West Bengal report indicated that 15% of families have difficulty in arranging two meals per day on a year-round basis. The calorie intake of the bottom 25% of the poor class declined to 1,624 kcal in 2004-2005, compared to the national norm of 2,100-2,400 kcal. In addition, it has been estimated that per capita calorie in-take of poor and rich in India in 2010 was 1,754 kcal for the poor and 2,819 kcal, respectively. Protein was 48 grams for the poor and 85 grams for the rich; whereas, fat was 29 grams for the poor and 71 grams for the rich. In spite of the fact that India achieved self-sufficiency in food at the macro level, 57% of pre-school children suffer from Vitamin A deficiency. According to FAO, the undernourishment of the total population was about 20% in 2012. With an annual increase in population of 16 million, resolution of the challenges of food security and malnutrition within a reasonable period of time will be difficult.

India's 2013 Food Security Bill policy recommended that action be taken to: 1) revamp small holding agriculture for sustainable higher production through local low-cost technologies and resources, 2) launch a watershed management program, 3) popularize growing fruit trees at homesteads and in upland degraded soil where regular field crops are not productive, 4) create more job opportunities in rural sectors through public works like rejuvenation of wells, water tanks, village roads, reforestation/social forestry, waste and crop residue cycling and vermicomposting and 5) subsidized food distribution programs. These programs should focus on school children of 6 to 14 years old and provide free food and health services for pregnant and lactating women (Saxena, 2013). In addition, as per the Food Security Bill, subsidize food at 2 rupees /kg for cereals for 67% of the total population. Selected national food security indices from the first year of the global food security evaluation process are presented in Table 5.1.

Table 5.1. 2012 Global Food Security Indices (GFSI) of fourteen nations[1].

USA	89.5	Denmark	88.1
Norway	88.0	France	86.8
Netherland	86.7	Austria	85.6
Switzerland	83.7	Canada	83.4
Finland	83.1	Germany	83.0
India	45.0	Pakistan	38.5
Nepal	35.2	Bangladesh	34.6

[1]www.socialissueindia.worldpress.com

Table 5.2 Three categories of the Global Food Security Index for seven nations in 2019.

Nation	Availability	Affordability	Quality/Safety	Index
Singapore	83.0	95.4	79.4	87.4
U.S.A.	78.3	87.4	89.1	83.7
Japan	71.0	82.4	76.7	76.5
China	66.9	74.8	72.6	71.0
India	58.4	64.2	47.0	58.9
Nigeria	45.8	50.4	50.7	48.4
Venezuela	32.2	15.8	66.9	31.2

Since the entire population of India is part of the food security issue, the initiative involves an annual expenditure of about one trillion rupees per year. This comes from the central budget. Presently, under PDS (Public Distribution System), there are nearly 450,000 fare price shops distributing commodities like rice, wheat, sugar, oil, pulses, kerosene etc. These are worth 300 billion ru/year and distributed to 160 million households as a supplemental supply of essential food. This, however, is not a permanent solution.

True food security is neither possible nor economically feasible, without generating national wealth through self-help groups, increasing agricultural production and creating employment opportunities on a priority basis. The issues of poverty, hunger and unemployment are closely related in rural and urban areas. They result in suffering related to food acquisition, but also for shelter, clothes, medicines and education for children. In 2004-2005, more than 300 million or 28% of India's population lived in poverty. Poverty is most common in MP, Bihar, Odisha, Jharkhand and UP. It exists mostly among scheduled castes (SCs) and scheduled tribes (STs) in rural areas where poverty is estimated to be between 64% to 76%. Social protection measures such as a subsidized food supply and employment generation programs can help to increase the purchasing power of poor people and reduce the poverty level. The lack of sustainable high agricultural productivity is the key issue that results in poverty and hunger. Ecosystem-based agriculture is essential for achieving the vision of attaining food security, and zero-hunger.

The overall well-being of India is dependent on its agriculture. It contributes to about 17% of India's GDP and provides employment for 60% of its population. Success depends on a favorable monsoon. Agriculture failure is often the result of erratic rainfall, prolonged drought, flood, deterioration of soil health and poor ecological harmony. In recent years, this has involved temperature increases, unscientific soil-crop water management and other improper agronomic practices. In 2020, India could attain self-sufficiency in food, particularly in cereals and grains; but pulses, oil seeds and sugar production lagged behind, resulting in the need to import these commodities. The lack of proper knowledge, awareness, training of farmers about site-specific technology in agriculture and resource management, including soil health, are the main constraints that affect the inadequate performance of agriculture (Bhalla *et al.*, 2001). The Green Revolution resulted in a steady 3.2% annual increase of food grain production. The growth rate, however, declined during the past two decades by slightly more than 1% resulting in serious concern about maintaining security under PDS (Ittyerah, 2013).

Maintaining food security by importing food is a false proposition. Development of ecosystem-based agriculture is the key factor for successful socio-economic and welfare programs of India. Food production in India was increasing marginally at 252 million tons in 2016 and targeted for 270 million tons by 2017. Globally, India is the second largest producer in wheat, rice, fresh vegetables, sugarcane, groundnuts, lentil, garlic, cauliflower, and broccoli. It is within the top five producers of cotton, jute, tea, fruit, vegetables, chicken, milk, and beef. About $39 billion of agricultural products are exported annually to South East Asia, Europe and the Americas. The growth rate of rice and wheat was slightly less than 5% from 1992 to 1997 and declined to about 2% from 1997 to 2002 and 1% during the 10th Plan of 2002-2007. India's rice yield is the lowest among rice producing nations. Wheat yields, however, are relatively high, and near the global average.

New policies are needed to improve agricultural productivity, increase food security and attain zero-hunger (Jha, 2007). The evidence indicates that conventional agriculture with its external inputs of synthetic fertilizers, synthetic pesticides, high-yielding varieties and genomic seeds is not adequate to obtain food security. The technologies of ecosystem-based agriculture using local resources, soil health building and crop-diversity properly integrated with the dynamics of climate, socio-economic conditions of farms and markets is required for a positive future of agriculture in India. The same is true on a global basis. New policy must be designed to achieve the following seven characteristics: 1) improved post-harvest technology, 2) adequate storage and warehouse facilities 3) market linkage for farmer profitability, 4) integrated agronomic and horticultural crops, livestock and fish systems, 5) soil health renovation, maintenance and enhancement, 6) balanced utilization and conservation of local resources and 7) enhanced rural infrastructure, including sanitation, drinking water and cyber technologies.

There needs to be mandatory training for Indian farmers. These programs must include instruction on the topics included in the recommended policy innovations. In addition, the training must provide education and demonstrations of the ways to preserve and use the 500 million tons of local organic resources generated on an annual basis. In China, about 700 million tons of crop residues are generated every year. Of this amount, about 31% is left in the field, 31% used as animal feed, 19% used for bio-energy generation and 15% maintained for soil health enhancement (Jiang *et al.*, 2012). In addition to compost and manure, these system residuals can be preserved and used for improving soil health and crop

137

productivity. Wherever possible, farmer-to-farmer education (an early adopter farmer serving as the teacher) and on-farm demonstrations should be used. Because of the dynamics of agriculture, cyber technologies can be important when and where the technologies are available.

In addition to enhanced agricultural productivity as a solution to zero-hunger, agriculture needs to strive for a blend of improved food quality, nutritious food, soil health and environmental quality at the local level. This must be designed to benefit the general welfare of people and animals. To achieve this, it must be recognized that 60% of the farmers practice agriculture on small marginal farms on degraded, arid, rainfed and hilly areas where conventional farming practices are difficult and often do not produce the desired results. This mandates that the following question be resolved. Should the conversion of conventional agriculture go directly to organic agriculture or should the transition be to ecosystem-based agriculture? Before answering this question, it is necessary to review five characteristics of Indian agriculture:

1) There are about 127 agro-climatic zones in India. These are based upon topography, climate, soil types, variation of temperature, humidity, rainfall, sunshine and biological diversity. This means that cropping patterns, crop distributions and production policies have to be designed for the best ecological fitness of each zone.

2) More than 80% of India's agriculture is achieved through conventional farming technologies. This includes most of the grains. A majority of the medium and large-scale farmers depend on synthetic chemicals and do

very little cycling of system residuals. These farms need to engage in a seven-year transition to ecosystem-based agriculture.

3) This will require farmer-to-farmer education programs and on-farm demonstrations. A short-term goal is to have at least 50% of the system inputs from local biological resources. Detailed electronic recording keeping is an important aspect of medium and large-scale ecosystem-based agriculture.

4) Small and marginal farmers have very limited purchasing capacity for system inputs such as hybrid seeds, fertilizers and pesticides. Many of them farm degraded land under arid to semiarid conditions. These farmers need to transition to a system of ecosystem-based agriculture focusing on the use of IPM, biopesticides, bio-stimulants, green manuring and organic mulching. The seven-year transition program for this system must also include farmer-to-farmer education and on-farm demonstrations under degraded, arid, rainfed and hilly conditions. These are areas where conventional farming practices are difficult and often do not produce the desired results. The program must include a focus on soil health and the basics of recording keeping.

5) Most tribal farmers that cultivate terraced lands and the small farmers of coastal, waterlogged and dry land belts frequently do not use external synthetic inputs. Successful

ecosystem-based agricultural practices can be designed specifically for these systems. The demonstration of ecosystem-based technology should be done by local leaders as part of farmer-to-farmer education programs.

Since the 1990s, there has been a significant decline in per capita availability of grain to 445 grams/day, compared to 469 grams/day in 1960 (Saxena, 2013). In addition, there have been declines in the condition of soil structure, soil fertility, overall soil health, and quality of agricultural products. This has taken place at a time of increasing input prices, economic instability of farmers, and poor health of rural people; especially children and women. Development of progressive site-specific education programs for Indian farmers is key to transitioning agriculture to a system designed to facilitate food security and zero-hunger.

Recommendations for resolving the current situation include aggressive planning by farmers concerning what to grow, how to manage the crop, site selection, site-specific water management, marketing through self-help groups, farmer clubs, co-operatives, contract farming and direct sale of products in local markets and urban chain stores. These have highly significant potential for appropriate change. No single tactic is likely to achieve the overall goal. This is a key reason why ecosystem-based agriculture is imperative for obtaining food security and zero-hunger. It must be remembered that Albert Einstein said, *you cannot solve a problem with the same kind of thinking that created the problem*. Appropriate change in agriculture requires change in policy. The U.S.A. Sustainable Agriculture Research and Education Program (SARE) consists of innovations that need to be considered in the next generation of India's agricultural policy. Its policies and funding decisions are the results of the involvement of meaningful dialogue among farmers, scientists, technologists, educators, government representatives, not-for-profit advocacy organizations and private sector business representatives. The policy needs to include organic agriculture as a special category of ecosystem-based agriculture.

During the last 10,000 years, agriculture has evolved from systems of ancient to traditional, conventional, ecosystem-based and organic systems. Ecosystem-based and organic agriculture involve a holistic approach, like that of Aristotle, where structure and process result in repeatable patterns. In many cases this pattern should involve a simple and optimistic living style, a way of moderate life in close harmony with nature, the environment and health of humankind.

The fundamental characteristics of organic farming system form a hexagon with the following six pillars (Fig. 5.1):

1) Soil health: maintenance, renovation or enhancement, including fertility, pH, water stable aggregates, nitrogen mineralization potential and active carbon, monitored periodically.

2) Ecological harmony and biological diversity of crops, beneficial organisms and integration with animals, non-cash cover crops and fish.

3) Site-specific technology for cycling with legume culture, crop rotations, mixed or intercropping, cover cropping, zero or minimal tillage, or alley cropping to maintain a green cover year-round through effective water management.

4) Use of various on farm organic manures, composts and vermicompost with or without

141

mineral additives, animal manures, green manuring, biological fertilizers and organic permissible biological stimulants.

5) Biological and environmental monitoring for IPM: decision-making that involves biological pest and infectious disease management in a manner that reduces ecosystem disturbance as much as possible while obtaining the desired crop productivity.

6) Farmers education, training, demonstration, market linkage, incentives and organic certification.

The fundamentals of ecosystem-based agriculture and organic agriculture are founded on the principles of justice and care. This mandates development of an optimal relationship between the biotic and abiotic components of the ecosystem and the care necessary to protect community health and the general welfare for present and future generations. In the end, the laws of nature will control the system and an Anthropocene is not attainable. Current and future agricultural policies must be designed and monitored for sustainable long-term benefits for food security, environmental health and potential zero-hunger. Healthy soil, fresh water, biological diversity, clean air and human ingenuity are the five core elements for food security and zero-hunger. For a reasonable quality of life for the global population of about 7.8 billion people, these five core elements are essential.

Future policy must include appropriate lifestyles, environmental quality, soil health and the nutrition of women and children. This will mandate change. It must be based on the fundamentals of education, facilitation and persuasion. It must focus on ecosystem-based agriculture, which is already practiced, at least in part, by small and marginal farmers of 170 nations. It must be a grassroots movement designed to foster intergenerational equity. About 700 thousand farmers are engaged in certified organic farming, mostly for exporting organic products at premium prices. This is successful because of consumer demand in North America and Europe. For the short-term, the conventional agricultural production of grains will be required to meet global consumption needs. A significant number of conventional farmers, however, currently observe organic and ecosystem-based practices to determine which of the practices used in these systems can be successfully adopted on their farms. This is a relatively recent and very positive development.

On a global basis, organic farming is growing rapidly. The products are mostly fruit, vegetables, spices, meat, milk, eggs, fish, coffee, tea, cocoa, and even rice, wheat and millets. Japan is a major importer of organic soybeans from the U.S. For the most part, certified organic food is not available as food for reduction of hunger and malnutrition. Resolution of challenges associated with physical access to food, hunger and malnutrition are imperative for a healthy humankind. Globally, about 795 million individuals go hungry or are undernourished. About 780-million of these are in Asia. In India, 57% of pre-school children suffer from Vitamin A deficiency and pregnant women are affected by chronic iron deficiency and anemia. Further, about 33% of the people in India face food insecurity and protein-energy malnutrition, which is difficult to resolve with cereals and grains

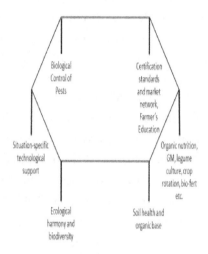

Six Pillars of Organic Agriculture

Achievement of zero-hunger would save the lives of 2.8 million children below the age of five. On an annual basis. The way to accomplish this is through the adoption of ecosystem-based and organic farming using indigenous crop-varieties rich in minerals and vitamins in addition to the products of conventional agriculture. The NPOP of India must develop and implement policy that facilitates the adoption of ecosystem-based and organic agriculture founded on local and site-specific natural resources and quality of life needs. This must include research and development in addition to education, facilitation and persuasion attributes.

The availability of organic resources, land, ecological fitness and socio-economic status of farmers must be integrated with ecosystem-based, organic and indigenous practices to provide a synergistic effect on soil health and sustainable food production. This can only succeed, however, if it includes active and collective participation by farmers, consumers, education institutions, non-government advocacy organizations and private business. One way to achieve this is through regional food security/zero-hunger councils, each consisting to two farmers and one representative from each of the other five categories (consumer, education institution, NGO, government and private business). This challenging, process will result in a highly significant trek towards local, national and global food security and zero-hunger. The system and its policies must be dynamic since it is certain that change will always take place.

Epilogue

The Epilogue consists of three parts. The first focuses on the overall nature of ecosystem-based agriculture. The second is a discussion of the impact of pandemics on food systems: with special reference to COVID-19. The Epilogue concludes with a recommendation to transition to an Ecological Society. The Ecosystem-based Agriculture Futures section is designed as a summary of the current state of the topic, with comments about its future role in global food security and progress on the trek towards zero-hunger. It includes specific policy recommendations for India and ends with a modification of Gandhi's *Seven Deadly Social Sins* and a mandate for a transformation from the current dominant Mechanistic World View to an Ecological World View. The section on pandemics includes brief discussions of several pandemics. It includes case studies of the impact of COVID-19 in the U.S. and how ecosystem-based agriculture is designed to reduce risks associated with pandemics and their potential impacts on food security and hunger. In concluding remarks, the authors sincerely hope that eco-literacy will contribute towards the evolution of an Ecological Society and overall enhancement of global quality of life.

Ecosystem-Based Agriculture Futures

Agriculture is humankind's greatest invention! Following its origin 10,000 years ago, it evolved from ancient to traditional and then conventional agriculture. Today, in a world of more than 7.8 billion people, ecosystem-based agriculture is the best route for obtaining global food security and zero-hunger. Ecosystem-based agriculture is founded on the laws of nature and science. It relies on everything around us, including air, water, soil, biological diversity and human ingenuity. It is all of what we see or perceive collectively from the known environment. It includes most of the attributes of what is referred to as bio-farming, eco-farming or natural farming. Organic farming is a special category of ecosystem-based agriculture. It requires third-party certification and prohibits the use of synthetic fertilizers, synthetic pesticides and GMOs. Ecosystem-based farming is designed in a way that does not require the use of synthetic fertilizers or synthetic pesticides. They can, however, be used on a limited basis in emergency situations.

Ecosystem-based agriculture is a highly honorable profession. It has a long history. The journey began in the environment of the eastern Fertile Crescent Mediterranean regions on both sides of the Nile river. The term environment includes all biotic and abiotic aspects of a system. Ecology is the study of the interactions among all physical, chemical, mechanical, biological, anthropocentric and pyric components of the system. Ecosystem-based agriculture was selected as the focus of this book because it is based on the dynamic equilibrium and harmony among biological diversity, soil health, organic nutrition, biological plant protection measures, ecological friendly cropping and animal production systems This integrates itself in the form of a hexagon as illustrated in Fig. 5.1. It focuses on sustainable and equitable production of healthy food, feed and fiber in a socio-economic and environmentally sound manner. Ecosystem-based agriculture is highly compatible with nature, the ultimate truth and power regulating our planet.

Ancient Vedic literature, the Upanishad (*ca* 2000-3000 BC), explains the cyclic action of "Pancha Bhuta". Under this hypothesis, all comes from and returns to Kshiti, (soil), Apa (water), Teja (light/energy), Marut (air), and Bhyom (sky known as the cosmic world). This ancient concept is very similar to the current scientific understanding of how the world is known to work.

For a farm to be healthy and reach an appropriate dynamic equilibrium, there must be a balanced integration of its plant, animal and microbial species. Symptoms of an unhealthy farm include poor soil health, pest/disease/weed problems and low productivity. Without proper balance, the system will not be healthy. Achieving this balance is the key factor for sustainable and equitable development. Ancient civilizations appear to have given priority to preserving the sanctity of natural resources for production of food or rearing animals and the overall health of their settlements.

At a meeting of the U.S.A. Sustainable Agriculture Research and Education Program Advisory Committee Meeting In Washington, D.C. in the early 1990s, the following question was proposed: *is it possible to farm sustainably without animals on the farm?* After about three hours of debate, it was declared that the answer is yes, but it is much more difficult than if animals are present. About ten years later, Professor Bird made an Extension visit to a Michigan apple farm with a serious tree decline and soil health problem. After about twenty minutes of discussion, the farmer said, *are you trying to tell me that the day I got into trouble was when the neighboring dairy farm went out of business?* Professor Bird responded, *yes.* That was when the orchard was forced to change from a manure-based soil fertility program to a synthetic fertilizer-based program.

By 7,000 B.C., agriculture had gradually shifted from the Nile sites to Far East Asia of Iran, Iraq, Palestine, Mesopotamia and Tripoli. Oats followed by wheat and other crops were cultivated. Rice was first grown around 3,000 B.C. in the low land of Indo China. The steppe regions of Russia, river sides of the Danube in Europe, and lake ridges in Switzerland were the sites for early oats and wheat. Two types of jower (sorghum) were ecologically suited for Italy. About 1,000 B.C. during the Vedic Period in the Indian sub-continents. Wheat, rice, sugarcane, gram, sesame, mustard, banana, cotton and vegetables were successfully grown on both sides of the Ganges in the Indus Valley. Cattle farming and agriculture were prestigious professions in Roman times. The use of canal irrigation at Mehergarh, Baluchistan existed around 6,000 B.C. Archaeological evidence indicates that animal manure, oil cakes, green manuring, wood ash were used for crop nutrition. These were included in the Rig Veda, Parasarasamhita about 1,000 B.C. and Kautilya's Arthasastra around 322-332 B.C.

Innovative and progressive traditional farming practices were used by Roman farmers around 510-523 B.C. In the Senegal river valley of the Congo, ancient farmers used local organic resources. In China, agricultural settlements were present in 500 B.C. By the 16th Century, the Iroquis and Huron tribes of North America cultivated maize, potatoes, beans, squash and sunflowers. Corn was grown in Ontario, Canada by 500 AD. The pattern of development of ancient agriculture and food production suggests there was a philosophical, ethical, and holistic approach to sustainable food production technology based on the nature of local ecosystems. It appears that special attention was given to the maintenance of natural flora, fauna and soil microbial populations. This is not common in conventional agriculture. It must be recreated to overcome the negative aspects of climate change, topography, demography, destruction of natural resources and burgeoning pressure of human population. In fact, today, a true steady state of rhythmic vegetative eco-climax is most likely only possible in very isolated terrestrial and aquatic eco-systems. The current demand of food security, cropping patterns, diversification and distribution exceed the threshold for optimal functioning of conventional systems. This results in soil, water, atmosphere and agricultural production degradation. Since the beginning of ancient agriculture, the human population has increased while land mass remains the same. In his book, *Half Earth*, E. O. Wilson proposes that 50% of the earth's surface must be maintained as a natural environment for humankind to have a high quality of life.

Conventional agriculture evolved shortly after the discovery of Liebig's 1840 Law of Minimum and continued to expand through the Green Revolution of the 1960s. During the last 40 years, soil health, agricultural production and air, water and light quality have deteriorated in many regions. Farming systems, therefore, must be designed to serve both current and future needs. They must be ecosystem-based: with special reference to short-term food security and long-term ecosystem health. This not only requires a balanced system, but most likely a two-system approach where some of the land is dedicated to soil health renovation and biological diversity enhancement. If soil health is lost, everything is endangered. Unfortunately, no single system can resolve this issue. Agricultural systems must be regional and site specific. Animals must play a pivotal role in supplying organic materials and cycling their residuals. Farming systems decline when animals are replaced with synthetics, machines or cyber technologies.

Conventional agriculture is practiced widely because of the perception of higher yields compared to other farming systems, well-defined fertilizer protocols and readily available synthetic insecticides and fungicides. Professor Bird questions the higher yield perception. He works with both ecosystem-based farmers and organic agronomic crop farmers that obtain yields equal to those of their region's conventional farmers. Ecosystem-based farming is practiced by about 13% of the farmers in Australia, 8% in North America, 9% in Asia, and less than 1% in India. This indicates that there is considerable
room for improvement. In Australia, most dry lands are in pastoral systems with dairy, poultry, hogs, or fruit orchards managed under ecosystem-based or organic practices. There are, however, many constraints to the adoption of ecosystem-based agriculture in the developing nations of Asia, Africa and Latin America. These include the: 1) absence of farmer incentives, 2) undeveloped domestic and consumers markets at the local level, 3) lack of proper farmer education and training, 4) inadequate insect pest and infectious disease plant protection research suitable for use in ecosystem-based systems and 5) comparative low

yields during the transition process or until the technologies are mastered. Ecosystem-based agriculture is attractive and should be strongly encouraged for the development and maintenance of a healthy agriculture, high rural employment and overall quality of life. It must be stressed, however, that a transition period is usually essential. There may also be times when synthetic fertilizers and pesticides are required on an emergency basis to save a crop. This is the reason the late Dick Thompson of Boone, Iowa, U.S.A., who was one of the original Rodale Organic Farmers, switched to ecosystem-based agriculture and founded the Practical Farmers of Iowa.

Healthy soil is an imperative for global food security and zero-hunger. The system of chemical fertilizers developed for conventional agriculture ignores soil biology and is not sustainable. The same is true for synthetic chemical-based pest and infectious disease management. Many soil health friendly technologies have been developed and validated in the past 30 to 40 years. Many of these practices are based on the age-old proven traditional systems of ancient and traditional agriculture.

To achieve global food security and zero-hunger, the six pillars of ecosystem-based agriculture must function in an optimal manner. This involves soil health and nutrition, ecological harmony, production technology support and integrated pest management. It must be remembered that education, facilitation and persuasion are key elements of the ecosystem-based agriculture transition process. In addition, new policies, scientific discoveries and technologies are needed. This is an arduous and challenging task. It requires a holistic and ethical approach that includes real-time environmental and socio-economic monitoring. At the onset, it must be remembered that the organic residuals form the basis of an ecosystem-based enterprise's future. In 1989, the senior author of this book stated that this process is governed by the laws of science and nature, as per the Roman maxim, *Nuisance is the mother of vegetation.*

Food Systems and Pandemics

Late in 2019, a serious infectious disease caused by a previously unknown double stranded ribonucleic acid (RNA) corona virus was detected in Wuhan, China. As with all viruses, it is a chemical messenger composed of nucleic acid and protein. When the virus enters the cell of a living host, it sheds its protein coat and the nucleic acid takes over the metabolic machinery of the cell to produce virus particles at the expense of the living host. This process is detrimental to the host's over-all physiology, resulting in an infectious disease. The new disease was quickly named COVID-19 (**Co**rona **Vi**rus Infectious **D**isease 2019) by the World Health Organization (WHO). Both the virus and the associated infectious disease spread rapidly on a global basis, resulting in a pandemic.

Pandemics occur periodically. Both Professor **Chakraborty** and Professor Bird were 18-years-old during the Asian flu pandemic of 1957 and Professor Bird's father was 18-years-old during the Spanish flu pandemic of 1918. One of the most significant pandemics ever was the Bubonic Plague. It is referred to as the Black Death. This infectious disease is caused by the bacterium, *Yersinia pestis*. This bacterial pathogen is vectored (transmitted from a diseased individual to a healthy individual) by the rat flea, *Xenopsylla cheopis*. Rats (*Ratus* spp.) are the primary host for the bacterium. Bubonic Plague pandemics occurred periodically, often separated by centuries. Molecular science indicates that humans have been impacted by this bacterial infectious disease for several millennia.

The first recorded Bubonic plague pandemic began in 541 A.D. It was known as the Plague of Justinian and lasted about two centuries. It is estimated to have resulted more than 50 million human fatalities at a time when global human population is believed to be only several hundred million. The next Bubonic plague pandemic was facilitated by the long-distance boat transportation system of the era. This infectious disease was brought to Europe in 1347 when boats from the east, with plague-infected individuals docked in Italy. Most of the passengers were dead and many others had readily visible symptoms. In 2020, COVID-19 was brought to the U.S.A. by infected airplane passengers arriving on international flights from China and Europe.

Bubonic plague still exists today in many diverse locations. It is managed through proper diagnosis. This includes flea and rat population reduction, general sanitation practices and antibiotics. During Soviet times, risk of plague in Central Asia was significant enough for the U.S.S.R. to have a separate branch of government for plague management. This employed thousands of individuals. As of the publication of this book, neither the length of the COVID-19 pandemic or its final global human mortality number are known. In addition, the chronic health issues and socio-economic impacts associated with the pandemic are unknown. As with all open, complex non-linear processes, COVID-19 is outside of the limits of complete control by humankind. It must be managed through implementation of a comprehensive and dynamic infectious disease management process.

COVID-19 is an example of why it is important to preserve the sanctity of nature and the environment for the welfare of humankind and all of the species of our planet (Quarles, 2020). A key question is how and why are COVID-19 and pandemics in general related to agriculture and food security. In the early days of the COVID-19 pandemic in the U.S.A., numerous cases occurred in large meat processing and packaging enterprises associated with conventional agriculture. On a short-term basis, this was resolved with strict policies related to quarantine, social distancing and construction of plastic barriers between workers. Masks and gloves were required at all times. In India, COVID-19 resulted in a policy that closed urban factories. This had a negative impact on a significant number of employees.

Many of these individuals were from rural areas and responsible for sending money to their local villages in support of their elders. Without these resources, poor marginal farmers practicing traditional agriculture, were worse off than usual (Peer, 2020). They were at risk of losing their health and land. This has the potential to result in increased food insecurity and hunger. Ecosystem-based agriculture is a process uniquely designed to assist in improving the overall quality of life of high-income and low-income nations during a pandemic and the associated recovery period.

Modern western medicine is based primarily on diagnostics, medications and mending physical structures in the human body. It is well known that viruses undergo mutations and new medications are required to keep ahead of the dynamics of viral pathogens. Holistic medicine provides a process that focuses on nutrition, personal health and active lives. In many respects it is similar to the fundamentals of ecosystem-based agriculture. Much remains to be learned about the role of nutrition in relation to risk of infectious diseases.

The World Health Organization has indicated that in general, a diversified diet including foods such as broccoli, garlic, ginger, spinach, sweet potato, yogurt, almonds, turmeric, fish oil, banana and egg-yoks can be important in the development of the human immune system. These are especially important in the diets of children and pregnant women. They are foods that are often common in ecosystem-based agriculture and frequently not part of local conventional agriculture systems. In addition, many indigenous herbal plants can be integrated into ecosystem-based agriculture. For optimal knowledge about human health and associated infectious and non-infectious diseases for the future, it will be necessary to utilize an overall ecological systems research and education approach. It must be a dynamic system that allows for a comprehensive understanding of the components and their interactions within local and regional ecosystems.

When Professor Bird's 2020 Michigan State University undergraduate class entitled, *Pests, Society and the Environment* was requested to design a management system for an infectious disease, they converted a 1978 integrated pest management model into a dynamic ecosystem-based infectious disease management monitoring system (Fig. 6.1). These young scholars indicated that this type of holistic systems approach is essential for transitioning to an Ecological Society, as recommended by Magdoff and Williams in their 2017 book entitled, *Creating an Ecological Society: Toward a Revolutionary Transformation.*

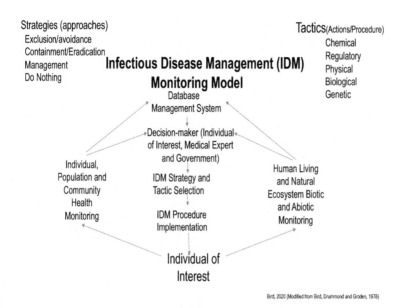

Figure 6.1 Conceptual model for infectious disease management.

Ecological Society

In 1993, Bird and Ikerd indicated that an appropriate blend of traditional, conventional, ecosystem-based and organic agriculture will be essential for achieving the challenges of food security and zero-hunger. In 2008, Ikerd enhanced this concept in his book entitled, *Small Farms are Real Farms: Sustaining People through Agriculture.* In Amanda Little's 2019 study of ecosystem-based, conventional, organic and factory food systems, she concluded by saying: *Three decades from now, if we are going to have a healthy and secure food supply, it will require a much larger, more deliberate network of participants, nationally and globally.* In addition, she indicated that when this is achieved, it will be in concert with the Piscatawayan American Indian motto of Pemhakamik Menenachkhasik, *The Whole World a Garden.* In their 2006 article entitled, *Future Potential for Organic Farming: A Question of Ethics and Productivity,* Kirschenmann and Bird indicated that the future paradigm for agriculture:

- *must meet the requirements of an exploding human population, in the face of entrenched poverty in a post-fossil fuel era,*
- *must restore the ecological health and natural resources on which agriculture depends, while the climate is changing and when global society insists that food is a human right, while increased infectious diseases require that we attend to the ecological ramifications of human activities and*
- *farmers must retain a sufficient share of the value of their productivity to be economically viable!*

To assure the goals of food security and zero-hunger, Kirschenmann and Bird modified Gandhi's *Seven Deadly*

Social Sins and recommended the following as a starting point:

- *food and farming systems must be regenerative in nature and based on cooperating partnerships and ecological independence,*
- *food and farming systems must be based on family enterprises cooperating to maintain vibrant communities and foster intergenerational equity,*
- *food and farming systems must generate appropriate wealth through work and*
- *food and farming systems must foster commerce inoculated with morality and politics with principles.*

This represents a starting point for the development of future healthy human living environments, sustainable natural resources and appropriate policy for individuals, communities, regions and the nations of our planet. Magdoff and Williams went a step further and recommended an ecological society in their 2017 land-mark treatise. This goal mandates a transition from the current dominant mechanistic worldview to an ecological worldview! To achieve this change, eco-literacy must be increased to a level where humans communicate with each other using the language of ecology!

Literature Cited

Agbonlabor, M., A. Aromolaran, and V. Aiboni. 2003. *Sustainable soil management practices in small farms in southern Nigeria, a poultry-food crop integrated approach.* J. Sus. Agric. 22:51-62.

Aina, P., R. Lal and G. Tailor. 1976. *Soil and crop management in relation to soil erosion in the rain forest regions of western Nigeria.* National Soil Conservation Conference, Lafayette, IN, USA.

Acharya, C., S. Bishoyi and H. Yaduvanshi. 1988. *Effect of long-term application of fertilizers and organic manures and inorganic amendments under continuous, cropping on soil physical* properties. Indian J. Agric. Sci. 48:509-516.

Aubert, C. 1972. *Agriculture Biologique Courier du Livre,* Paris, p. 251.

Anonymous. 2016. *Report, State Policy of Organic Farming.* Sikkim Organic Mission, FS & AD and HCCD Depts., Govt. of Sikkim, Today, East Sikkim, India.

Baishya, A., B. Gogoi, J. Jajarika, J. Hajarika, A. Bora, A. Das, M. Barah and P. Sutradhar. 2017 *Comparative assessment of organic, inorganic and integrated management practices in rice (Oryza sativa) based cropping system in acid soil of Assam(India).* Indian Journal of Agronomy 62:118-126.

Baishya, A., G. Gogoi, J., Hajarika, M. Borah, A. Borah, A, Rajbangsi, P. Deori and P. Sutradhar. 2015. *Effect of continuous cropping and integrated nutrients management practices on soil properties and yield of Rice: Rice in acid soil.* Indian J. Agron. 60: 493-501.

Barr, C. 1993. 1990 Countryside Survey, Main Report, Dept. Of Environment. East Coto, UK.

Bartwal, A., R. Mall, P. Lohani, S. Guru, and S. Arora. 2013. J. Plant Growth Reg. 32:216-232.

Basso, B., G. Shuai, J. Zhang and G. Robertson. 2019. *Yield stability analysis reveals large-scale nitrogen loss from US Midwest.* Scientific Reports. Nature. Open Access No. 5774.

Basu, R. 2008. *Strategy for ecologically sustainable agriculture: crop-livestock integrated holistic farming systems for small marginal farmers of India,* National Conf. Application Chemicals and Biological Technologies. Jadavpur University, Kolkata-32.

Basu, K. 2008. *When does growth trickle down to the poor: the India case study.* Cambridge J. Econ. 32:461-477.

Beanchi, F., C. Booij and T. Scharntke. 2006. *Sustainable pest regulation in agricultural landscapes: a review on landscape composition, biodiversity and natural pest Control.* Proc. R. Soc. 273:1715-1727.

Bhalla, G., P. Hazell and Kerr. 2001. *Prospects for India's cereal supply and demands to 2020.* Food, Agric. and Environ. Dis. Paper 29., Int. Food Security Institute, Washington, D.C.

Bhattacharya, D. 2011. *Paddy field soil conservation.* Agric. Sci. 2:341-346.

Bhattacherjee, P. and G. Chakraborty,G. 2005. *Current status of organic farming in India and other countries. Indian J. Fert. 9:111-123.*

Bhattacherjee, S. and G. Chattapadhyay, 2006. *Effect of vermicomposting on the transformation of trace elements in fly ash.* Nature Cycle Agro-Eco. Sys. 75:223-231.

Bhattacherjee, R. 1987. *Effect of Mussorie rock phosphate with phosphor-bacterin-organic additives on growth and yield of crops in wheat-based cropping system.* Ph.D. Dissertation. Dept. of Agronomy, Visvabharati Univ. West Bengal.

Bird G. and B. Basso. 2018. *Relationships among thermo-stability, soil health and potato tuber yield.* Proc. Great Lakes Agric Exposition.

Bird, G. and J. Smith 2013. *Observations on the biology of organic orchard soils.* Acta Hort. 1001:287-293.

Bird, G., L. Wernette. and L. Lott. 2012. *Potato nematodes; a farm guide to management.* Department of Entomology, Michigan State University. USA. 24 pp.

Bird, G. and J. Ikerd, 1993. *Sustainable Agriculture: A Twenty-first Century, System.* Ann. Am. Acad. Pol. Sci. 529:92-102.

Bird, G., T. Edens, F. Drummond and E. Groden (1990) *Design of pest management systems for sustainable agriculture,* (in) *Sustainable Agriculture in Temperate Zones,* C. Frances, C. Flora, and L. Kieg (eds) John Wiley & Sons.

Bird, G. 1996. *Sustainable agriculture and the 1990 Farm Bill.* Proc. Philadelphia Soc. Pro. Agric., Philadelphia, PA. USA.

Bisoyi, R. 2006. *Scope of organic farming in Eastern India.* Proc. National Seminar, Organic Farming: Resource of

Traditional Agriculture, pp. 25-27, Jadavpur University, Kolkata.

Biswas, N. 2015. *Accessing soil quality indices for sub-tropical rice-based systems in India.* Soil Research. DO10.1071/SR14245.

Biswas, R. 2014. *Organic agriculture,* pp. 141-165 (in) *Advances in Extension Education and Rural Development,* D. Dasgupta (ed), Agro. Bio. Vol. 1. India Pub. Jodhpur-342002.

Boehnert, J. 2019. *Design Ecology Politics*: Towards the Ecocene. The Design Journal 22:905-909.

Bouis, H. 1993. *Measuring the sources of growth and yields: Are growth rates declining in Asia?* Food Res. Inst. Studies, 22(3).

Brooks, F., 1995. *The Mythical Man Month: Essays in Software Engineering* (3rd ed). Addison-Wesley Pub. Co. Reading, Mass.

Bryden, J. 2002. *Rural development indicators and diversity in the European Union.* Conf. of measuring rural diversity, Washington, D,C. 15 pp.

Bugg, R. and C. Pickett. 1998. *Enhancing biological control: habitat management.* Univ. California Press, 123 pp.

Bulgari, R., G. Cocctta, A.Trevellini P. Vernieri and A. Ferranti. 2015. *Bio stimulants and crop responses: a review.* Bio. Agric. and Hort. 31:1-17.

Campagnoni, A., R. Pinton and R. Zanoli. 2000. *Organic farming in Italy,* (in) *"Organic agriculture in Europe: current status and future prospects of organic farming in 25 European countries.* S. Graf and H. Willer (eds). So"l Bad Durkheim.

Carson, R. 1962. *Silent spring.* Houghton Miffin. Boston, U.S.A. 368 pp.

Cato the Censor. 1933. *Records of civilization: on farming.* E. Brehaut (trans). Colombia University Press, USA.

Chakraborty, D. 1999. *India: an archaeological history,* New Delhi.

Chakraborty, T. 2014. *Ecological perspective of modernity in agriculture and sustainable eco-farming systems.* pp. 41-74 (in) *Advances in extension education and rural development.* Agro. Bios. India Pub. Jodhpur.

Chakraborty, T. 1992. *Land disposal of agricultural wastes: A low cost technology for soil conservation and sustained productivity,* (in) *Environment conservation and small-scale farming,* Hurn and Kebede (eds). ISCO Conf., Ethiopia and Kenya. Geographica Bernesia, MO, USA.

Chakraborty, T. 1973. *Nature of competition between weeds and rice under dry land conditions.*

Chakraborty, T. and S. Sen. 1967. *Mineralization of carbon and added organic matter in soil under laboratory conditions.* J. Indian Soc. Sci.15:153-155.

Chatterjee, A. 2005. *Ecological farming and NRM.* Solution Exchange 10, New Delhi: Food and Nutrition

Chaudhri D. and A. Dasgupta. 2011. *Agriculture: a study of Punjab.* Routledge Library Editions.

Chen, A. and S. Scott. 2014 *Chinese path in developing organic agriculture: rural development.* Ph.D. Dissertation. Waterloo Univ. Ontario, Canada.

Ciccarese, L. and V. Silli. 2016. *Role of organic farming for food security: local news with global view.* Future Food: J. on Food, Agriculture and Society 4:1.

Clements, F. 1936. *Nature and Structure of the Climax. Ecology* 24:252-284.

Crutzen, Paul. 2000. *Anthropocene.* International Geosphere-Biosphere Newsletter. Igbp.net.

Dale, T. and V. Carter. 1955. *Topsoil and civilization.* Univ. of Oklahoma Press, Norman, OK. 270 pp.

Dasgupta, D. 2014. *Use of traditional knowledge base for attaining food and ornamental security,* Adv. Ext. Ed. and Rural Dev. Agro. Bio., Jodhpur, India.

Dasgupta, M.K. 2005. *Constraints in suitability of organic farming in India. P. 12 (in) Solution Exchange, New Delhi: Food and Nutrition Security Community,* UN Team India..

Dasgupta, S., B. Laplante, S. Murray and D. Wheeler. 2011. *Exposure of developing countries to sea level rise and storm surges.* Climate Change106:567-579.

Dash, M. 1994 *Fundamentals of Ecology,* McGraw Hill Ltd., New Delhi.

De, A. 2014. *Biodiversity and natural resources management in the perspective of climate change,* (in) Adv., Ext., Ed. and Rural Dev., D. Dasgupta (ed). Agro. Bios. Jodhpur, India.

De, A. 2010. *Environmental Chemistry,* (7[th]ed) New Age International Ltd., New Delhi.

Derspck, R. 2001. *In sustaining the global farm.* pp. 248-254. (in) ISCO. Conf. Proc. D. Stott, R. Mohtar and G. Steinhardt (eds). Purdue Univ. W. Lafayette, U.S.A.

Dey, G. and S. De. 2015. *Bharatbarsher Itihas ancient history of India: prehistoric to ancient mid-age.* Pragatishil Pub. Kolkata, India, pp. 133-315.

Dhar, N. 1959. *Land fertility improvement.* J. Indian Soc. Soil Sci. 48: 509-516.

Diamond, J. 2005. *Collapse: how societies choose to fail or succeed.* Penguim, Random House Press, New York, USA. 592 pp.

Dunger, W. 1983. *Tiere im boden* (3rd ed). Wittenberg, Lutherstadt, Ziemsen.

Dunning, J., B. Danielson, and H, Pullium. 1992. *Ecological processes that affect populations in complex landscapes.* OIKOS 65:169-175.

FAO. For citations dates after 1999, see fao.org.

FAO. 1999. *Global world of organic agriculture.* Food and Agricultural Organization, Nuremberg.

FAO. 1997. *Bulletin on organic agriculture.* Food and Agriculture Organization.

Flinn, J. and S. De Datta. 1984. *Trends in irrigated rice yields under intensive cropping.* Philippine Research Station: Field Crops Research No. 9.

Flint, M. L. and R. van den Bosh. 1979. *Introduction to Integrated Pest Management.* Springer Books. N. Y.

Forge, F. 1998. *Agriculture conservation in Canada.* Rept. Sci. Tech. Div., Parliamentary Research Branch, Canada.

Gaje-Wolska, J., T. Spizcwski and A. Graboska. 2013. *The effect of seaweed extracts on yield and quality parameters of broccoli (Brassica oleracea, var. cymosa) in open field production.* Acta Hort. 1009: 83-89.

Gaje-Wolska, J., K. Kowalczyk, M, Nowceka, K. Mazur and A. Metcra. 2012 *Effect of organic mineral fertilizers on the yield and quality of endive.* Acta Sci. Pol. Hort. Cul.11:189-200.

Geetha, H. and H. Shetty. 2001. *Induction of resistance in pearl millet against downy milddew disease.* Crop Protection 21: 601-610.

Geissert, D. and J. Rossignol. 1987. *La morphologia en la ordenacion des Paesajes rurales.* Concepto y premieres applications Mexico. INIREB.

Ghose, M. and M. Dasgupta. 1999. *Potentiality of natural control against Insects-pests (in) Frontiers of plant protection,* Vol. 2, K. Dasgupta and A. Datta (eds) Dept. of Plant Protection, Palli Siksha Bhavana, Visvabharati.

Glasstetter, M. and H. Leser H. 1987. *Arbeitbericht Uber vergleichende bodenfauna-unter suchungen im Jura als Beitrag zur Enforshung beiotischer kompart-imente in geoo kosystemen.* Regio-basilensis 28:183-197.

Gonzalez, A. and R. Nigh. 2005. *Small holder participation and certification of organic farm products in Mexico.* J. Rural Studies 21:449-460.

Grinnell. J. 1917. *The Niche-Relationship of the California Thrasher.* The Auk 34:427-433.

Grunder, M. 1985. *Erosion, conservation and small-scale farming.* Geographica Bernesia Walswarth Pub., MO USA.

Guinan, K., N. Sujeeth, R. Copeland, P. Jones, N. Obrien, H. Sharma, P. Proutcau, and J. Sullivan. 2013. *Discrete roles of extracts of Ascophyllum nodosumin enhancing plant growth and tolerance to abiotic-biotic stress.* Acta Hort. 1009: 127-136.

Gurr, G., S. Wratten and J. Luna. 2003. *Multifunction agricultural biodiversity pest management and other benefits.* Basic Appl. Ecol. 4:107-116.

Gut. L., C. Adams, J. Miller. P. McGee, D. Thompson. 2019. *Biological control in integrated management of deciduous fruit insect pests: the use of semiochemicals.* Researchgate.net. DO 10.19103/AS.2019.0046.027.

Haider, K., C. Ayyub, M. Peruez and H. Asad. 2012. *Impact of foliar applications of sea-weed extracts on growth and yield of potato (Solanum tuberosum L.).* Soil and Environmental Science 31:157-162.

Hallsworth, E. 1987. *Anatomy, Physiology and Psychology of Erosion.* John Wiley & Sons, NY. 176 pp. (CABI, N.Y. 185 pp.).

Halberg, N. 2006. *Global development of organic agriculture: Challenges and Prospects.* CABI. p 297.

Harold, L., G. Triplett and W. Edward. 1970. *Soil erosion and agricultural sustainability.* Agri. Eng. 51:128- 131.

Heider, M., C Ayyub, M. Pervej, H. Asad, A. Manan, S. Raja, I. Ashraf. 2012. *Impact of foliar application of seaweed extract on growth, yield and Quality of Potato (Solanum tuberosum, L),* Soil Env. 31:157-162.

Hill, S. and J. MacRae. 1984. *Organic farming in Canada.* Dept. of Entomology, Ecological Agric. Projects, MacDonald College of McGill Univ. Canada. Classics Pub. U.S.A.

Hill, S. 1975. *Agricultural insect pests of tropics and their control.* Cambridge University Press. UK.

Hoddle, M., J. Grandgirard, Petit., G. Roderick and Davies. 2006. *Grassy winged. sharpshooter, ko-ed first round, in French Polynesia.* Bio-Control News and Information 27:47N-62N.

Hurni, H. and T. Kebede. 1992. *Erosion, conservation and small-scale farming.* Geographica Bernesia, Walswarth Pub., MO, USA.

Hurni, H. 1983. *Soil erosion problems and soil conservation design for Didessa State Farm.* Soil Conservation Res. Project Field Reptr. Addis, Ababa,

IFOAM. 2006. *Organic agriculture and food security dossier.* International Fed. of Organic Agric. Management.

Ikerd, J. 2008. *Small Farms are Real farms: Sustaining People through Agriculture.* Aces USA. Austin. 249 pp.

Ittyerah, A. 2013. *Food security in India: Issues and suggestions for effectiveness.* Theme paper, 57th Members Annual Conference, Indian Institute of Public Administration.

Ives, A., J. Klug and K. Gross. 2000. *Stability and species richness in complex Communities.* Ecol. Leu. 3:399-411.

Jha, P. 2007. *Some aspects of well-being of India's agricultural laborers in the context of contemporary agrarian crisis.* Ind. J. Labor Econ.

Jiang J. *et al.*, 2012. *Evaluation of cloud and watr vapor in CMIPS climate model using NASA's A-Train satellite observations.* J. Geophys Res. 117 No. D14.

Joshi, P., M. Wadhwani and J. Pradhan. 1968. *Ancient Indian history, civilization and culture.* New Delhi.

Jones, D. 1998. *Pepperoni butoxide- the insecticide synergist.* Academic Press, London.

Johnson, C. and W. Moldenhauer. 1979. *Effect of chisel versus moldboard plowing on soil erosion by water.* J. Soil Sci. Soc. Amer. 43:177-179.

Judson, S. 1968. *Erosion rates near Rome, Italy.* Science. 160: 1444-1446.

Kirkby, M. 1969. *Erosion by water on hill slopes,* pp 98-107 (in) *Introduction to fluvial process,* R. Chorley (ed). Methuen & Co. Ltd.

Kirschenmann,F. and G. Bird. 2006. *Future potential for organic farming: a question of ethics and productivity,* pp. 307-324 (in) *Developing and extending sustainable agriculture: a new social contract,* C. Francis, R. Poincelot and G. Bird (eds). Hayworth Press Inc., N.Y. 367 pp.

Klages, K. 1942. *Ecological crop geography.* Soil Science 54:1-74.

Klichestein, D. 2004. *Secondary metabolites and plant/environment interactions: a view from Arabiodopsis thaliana_tinged gases.* plant Cell Environ 27:675-684.

Koch, R. and F. Loeffler. 1876. *Investigations into bacteria. V. The etiology of anthrax based on the ontogenesis of Bacillus anthracis.* Cohns Beitrage zur Biologie der Pffanzen 2:277-310.

Kramer, Tyler. 2020. Illustrator, *Ecosystem-Based Agriculture: The Pillar of Global Food Security*. Honors College, Department Business and Economics, Michigan State University, East Lansing, U.S.A.

Kumar H., A. Shafaat and Z. Sunil. 2015. *Efficacy of bio agents and botanicals against brown spot disease*. Oryza, 52:54-58.

Kumar, A., M. Bautilan, P. Kumar, S. Kumar and S. Jee. 2012. *Food security in India: trends, patterns and determinants*. Ind. J. Econ. 67: 445-463

Leser, H. 1980. *Soil erosion measurement on arable land in North-West Switzerland*: Geography in Switzerland. Geo. Helvetica 5:9-14.

Linker, H., D. Orrs and M. Barbercheck. *Insect management on farms*. Centre for Environment Farming Systems (www-Cefs.nesu.edu.).

Little, A. 2019. *The fate of food: what we'll eat in a bigger, hotter and smarter world*. Penguin Random House LLC, N.Y. 340 pp.

Loch, R. 2004. *Soil conservation practice: in search of effect solution*. International Soil Conservation Organization Conference.

Maass, J. 1992. *The use of litter mulch to reduce erosion on hilly lands of Mexico*. (in) *Erosion, conservation and small-scale farming*, E. Hans (ed) Geo. Bernesia, MO, USA.

Magdoff, F. and C. Williams. 2017. *Creating an ecological society: toward a revolutionary transformation*. Monthly Review Press, N.Y. 387 pp.

Mahapatra, I. 1990. *Production constraints and future prospects of rice in Eastern India.* pp. 242-312 (in) Proc. Int. Sym. on National Resource Management for a Sustainable Agriculture, New Delhi.

Marsh, G. 1864. *Man and nature or physical geography as modified by human action.* University of Washington Press. 2003.

McNaughton, S. 1979. *Grassland-herbivore dynamics,* pp. 46-81 (in) *Serengeti dynamics of an ecosystem,* A. Sinclair and M. Norton-Griffiths (eds), Univ. Chicago.

Meadows, D., D. Meadows and J. Randers. 1992. *Beyond the Limits.* Chelsea Green Pub. VT, U.S.A.

Michalopoulos. S. 2015. *Commission of Organic Farming Food Security.* Euractive, Greece.

Miguel. Alfieri. 1999. *The ecological role of biodiversity in ecosystem agriculture,* pp. 19-31 (in) *Ecosystems and Environment* Vol.74.

Miller, G. 2007. *Environmental Science: Working with the World.* Cengage Learning. 436 pp.

Mishra, P. and S. Rai. 2013. *Use of indigenous soil and water conservation practices among farmers of Sikkim Himalaya.* Indian J. Traditional Knowledge 12: 454-464.
Literature Cited

Montgomery, D. 2012. *Dirt: The Erosion of Civilizations* (2nd ed.). University of California Press. Berkeley, CA.

Morgan, R. 1986. *Soil erosion and conservation.* Harlow: Longman Scientific Tech. Bul.

Murdoch, J., Y. Ward, and P. Lowe. 1992. *Rural sustainable development: a socio-political perspective on role of agriculture.* Circle for Rural European Studies Conf. Mediterranean Agronomic Institute of Chania, Crete.

Nadaf, A., S. Krishnan, and K. Wakte (2006). *Research communication*, University of Pune, Current Science Vol. 91.

Nasim A., L. Yeasmin, S. Chakraborty, S. Sinharay and P. Sathi. 2014. *Role of biodiversity in development,* pp. 177-182 (in) *Advances in Extension Education and Rural Development*, D. Dasgupta (ed) Vol. 1.

Nitant, H., C. Hazra and M. Datia. 1998. *Grasses for rain - water conservation, erosion control and development of degraded land,* (in) *Frontiers of grassland management and research in North- East India,* T. Chakraborty (ed), Visva Bharati, Sriniketan, India.

Norrie, J. and J. Keathley. 2006. *Benefits of Ascophyllum nodosu, marine plant extract application to Thompson seedless grape production.* Acta Hort.727:243-247.

Odum, E. 1971. *Fundamentals of Ecology* (3rd ed). Sounders, Philadelphia, USA.

Okigbo, B. 1995. *Major farming systems in the low savanna of SSA and the potential for improvement.* IITA/FAO Workshop, Ibadan, Nigeria.

Okigbo, B. 1981. *Evaluation of plant interaction and productivity in complex mixtures as a basis for improved cropping system design.* (in) *Proc. Int. Workshop on Intercropping (ICRISAT).* Hyderabad. India.

Okigbo, B. 1978. *Cropping systems and related research in Africa.* Occasional Pub. No. 1. Addis Abeba, Assoc. Adv. Agri. Sci. in Africa.

Paez, M. and O, Rodriguez. 1992. *Erosion, conservation assessment on arable lands in Venezuela,* pp 39-49 (in) *Erosion, conservation and small-scale farming,* H. Hurni and T. Kebede. Geo. Bernesia Pub. MO, USA.

Pal, B. 1982. *Environmental conservation and development.* Indian Env. Soc. New Delhi.

Paramaiyan, P., N. Helberg and J. Hermansen. 2009. *Organic agriculture in relation to food security of developing countries.* Workshop 11. Nordic Organic Conference, Goteborg, Sweden.

Patra, A., K. Mishra, L.Garnayak, J. Halder and S. Swain. 2017. *Influence of long term integrated nutrients management on productivity and soil properties in rice (Oryza sative) cropping system in an acid soil.* Indian J. Argon. 62:111-117.

Peer, B. 2020. *Taking Amrit home.* New York Times (August 2). p7.

Peters, G. 1977. pp 592-610 (in) Proc. Int. Symposium. on Nitrogen Fixation. Washington State Univ. Press.

Pimental, D., P. Hepperly, J. Hanson and D. Douds. 2005. J. Biosciences, (Oxford) 55:573-582.

Pimentel, D. 1995. *Environmental and economic costs of soil erosion and conservation benefits.* Science 267:1117-1128.

Pimentel, D. 1993. *Environmental and economic effects of reducing pesticides use in agriculture.* Agric., Eco. and Env. 46:273-288.

Pornpratansombat, P., B. Bauer and H. Boland, H. 2011. *The adoption of organic rice farming in north eastern Thailand.* J. Org. Sys. 6: 2011.

Prasuhn,V. 1992. *A geo-ecological approach to soil erosion in Switzerland,* pp 27-37 (in) Erosion, conservation and small-scale farming, H. Hurni and K. Tato (eds). Geo. Bernensia, Bern.

Prabhu, M., A. Kumar and K. Rajamani. 2010. Indian J. Agric. Res. 44:48-52.

Pretty, J. 1995. *Regenerating agriculture: policies and practices for sustainability and self-reliance.* Earth Scan Pub. Ltd., London.

Quarles, W. 2020. *Coronavirus and Ecology.* The IPM Practitioner 37:1-7.

Radhamani, S., A. Balasubramanian, A. Ramamoorthy and V. Geethalakshmi. 2003. *A review on integrated farming systems.* Agriculture Reviews 24:204-210.

Roger, P. 1995. *Biological N^2 fixation and management in wet land rice cultivation.* Fertilizer Research 42: 261–276. Kluwer Academic Pub. Netherlands.

Rogers, E. and F. Shoemaker. 1971. *Communications of Innovations: A Cross Culture Approach.* Free Press, N.Y. 476 pp.

Russell, E. 1965. *Soil conditions and plant growth.* Longmans Green & Co. Ltd., London

Samado, H., E. Guei and N. Nguyna. 2016. *Overview of rice in Africa.* Africarice.org.

Saxena, N. 2013. *Hunger, under-nutrition and food security in India.* Paper No. 44 (in) India Institute of Public Administration 66 pp.

Scoones, S. and L. Elsaesser. 2008. *Organic agriculture in China-current situation and challenges.* European Union Rep.

Sharma, S.K. 1988. *Effect of industrial organic wastes and lantana incorporation on soil properties and yield of rice.* Indian J. Agron. 33:225-226.

Sharma, H., C. Fleming, S. Rao and T. Martin. 2014. *Plant biostimulants: a review on the processing of macro-algae and use of extracts for crop management to reduce abiotic and biotic stress.* J. Appl. Phycol. 26:465-490.

Shen, H. Lu, X. Wen and S. Jian. 2007. *Degraded ecosystems In China: status, causes and restoration efforts.* Land Scape Ecol. Eng. No. 1-13.

Singh, H. 1990. *Management of soil and water resources for sustainable agriculture.* Proc. Int. Sym. Natural Resource Management for Sustainable Agriculture 1:40-68.

Stern, V., R. Smith, R. van den Bosch and K. Hagen. 1959. *The integrated control concept.* Hilgardia 29:139-154.

Swaminathan, S. 2019. *Dietary iron and anemia are weakly associated, limiting effective iron fortification strategies in India.* J. Nutrition 149:831-839.

Swan, L. 1964. *Beneficial insects.* 1st ed. 249 pp.

Tafesse, A. 1992. *Soil erosion and conservation in large scale mechanized farms in Ethiopia,* (in) *Erosion, conservation and small-scale farming,* H. Hurni and T. Kebede (eds) Geo. Bernesia Pub. MO, USA.

Teasdale, J. 2016. *Strategies for soil conservation in no-tillage and organic farming.* J. Soil Water Con, 71, Uganda Org. Agric. Policy. Sustainabledevelopment.un.org.

Tisdall, J. 1985. *Earthworm activity in irrigated red-brown earths used for annual crops in Victoria.* Aus. J. Soil Res. 23:291-299.

Upadhyay, R., K. Mukherjee and R. Rajak. 1984. *IPM systems in Agriculture.* Vol.3. Cereals, Aditya Books, Delhi.

Vaidyanathan, S. 2017. *An Indian NGO's unique idea to curb food waste: taking from wedding and giving to the poor.* Food and Ecology.

Von-Braun, J. 2005. *Feeding the hungry in Africa.* U.N. Chronicle.

Wessels, T. 2013. *The Myth of Progress: Towards a Sustainable Future* (Rev. and Exp. Ed.). Univ. New England Press. Hanover, NH. 155 pp.

Willar, H. 2010. Federation of Organic Agriculture Management. www.fibl.org.

Willar, H. 2008. *Organic farming: China status, market trends and accreditation.* www.fibl.org.

Willar, H. and J. Lernoud. 2014. *The world of organic agriculture statistics and emerging trends.* FiBL-IFOAM Report.

Wilson, E. 2016. *Half Earth: Our Planet's Fight for Survival.* Liveright Pub., 259 pp.

Wood, S. and A. Worham. 1986. J. Soil Water Con. 51:193-196.

Wright, R. T. 2005. *Environmental Science: Towards a More Sustainable Future.* Prentice Hall. Englewood Cliffs, N. J. 630 pp.

Xu, H. 2013. *Nature farming in Japan.* DOI 10,1300J144v03nol_01. 170 pp.

Zaltman and Duncan. 1977, *Strategies for planned change.* John Wiley and Sons, N.Y. 404 pp.

Zinn, H. (2005). *A peoples History of the United States,* Harper Perennial Modern.

Made in the USA
Middletown, DE
02 August 2021

45243043R00106